Putting Water Security to Work

Over the last decade, water security has replaced sustainability as the key optic for thinking about how we manage water. This reframing has offered benefits (including clear recognition of the link between humans, the environment and the right to water) and also posed challenges (the tendency in some quarters to interpret "security" solely in terms of geopolitical or economic "securitisation").

In this collection, the authors offer a radical repositioning of these debates updated to reflect the concerns of our post-pandemic world. The chapters in this volume examine several different themes including how water security articulates with locality and culture, how it operates across spatial scales and its moral/ethical resonances.

The chapters in this book were originally published in the journals *Water International* and *International Journal of Water Resources Development*.

Chad Staddon is a Professor of Resource Economics and Policy at the University of the West of England. His research focusses on the social, political and economic issues related to water services, particularly drinking water. Current projects include the historical geography of urban water systems around the world, water-energy trade offs in unconventional oil and gas operations and appropriate socio-technologies for achieving universal access to safe water (SDG6). He received his PhD in Geography from the University of Kentucky in 1996 for research on the political economy of water (mis)management in post-communist Bulgaria.

Christopher A. Scott is the Director of the Udall Center for Studies in Public Policy and Professor of Geography, Development & Environment at the University of Arizona. Scott is an interdisciplinary scholar focusing on water, energy and food policy. His research and engagement address the policy dimensions of global change (climate change and urban growth) with particular emphasis on water and energy security, climate adaptation, urban wastewater and water reuse, agricultural-urban water transfers and transboundary water resources. He has lived and worked in South Asia and Latin America. Scott received PhD and MS degrees from Cornell University.

Routledge Special Issues on Water Policy and Governance

Edited by:

Cecilia Tortajada (IJWRD) – Institute of Water Policy, Lee Kuan Yew School of Public Policy, NUS, Singapore

James Nickum (WI) – International Water Resources Association, France

Most of the world's water problems, and their solutions, are directly related to policies and governance, both specific to water and in general. Two of the world's leading journals in this area, the *International Journal of Water Resources Development* and *Water International* (the official journal of the International Water Resources Association), contribute to this special issues series, aimed at disseminating new knowledge on the policy and governance of water resources to a very broad and diverse readership all over the world. The series should be of direct interest to all policy makers, professionals and lay readers concerned with obtaining the latest perspectives on addressing the world's many water issues.

Legal Perspectives on Bridging Science and Policy
Edited by Mara Tignino, Raya Marina Stephan, Renée Martin-Nagle and Owen McIntyre

Virtual Water
Implications for Agriculture and Trade
Edited by Chittaranjan Ray, David McInnes and Matthew R. Sanderson

Flood Resilience of Private Properties
Edited by Thomas Hartmann, Willemijn van Doorn-Hoekveld, Helena F.M.W. van Rijswick and Tejo Spit

A River Flows Through It
A Comparative Study of Transboundary Water Disputes and Cooperation in Asia
Edited by Selina Ho

Putting Water Security to Work
Addressing Global Sustainable Development Challenges
Edited by Chad Staddon and Christopher A. Scott

Global Water Resources
Festschrift in Honour of Asit K. Biswas
Edited by Cecilia Tortajada and Eduardo Araral

For more information about this series, please visit: https://www.routledge.com/series/WATER

Putting Water Security to Work

Addressing Global Sustainable Development Challenges

Edited by
Chad Staddon and Christopher A. Scott

Routledge
Taylor & Francis Group
LONDON AND NEW YORK

IWRA

First published 2022
by Routledge
2 Park Square, Milton Park, Abingdon, Oxon, OX14 4RN

and by Routledge
605 Third Avenue, New York, NY 10158

Routledge is an imprint of the Taylor & Francis Group, an informa business

British Library Cataloguing-in-Publication Data
A catalogue record for this book is available from the British Library

ISBN13: 978-0-367-65019-3 (hbk)
ISBN13: 978-0-367-65020-9 (pbk)
ISBN13: 978-1-003-12744-4 (ebk)

DOI: 10.4324/9781003127444

Typeset in Minion Pro
by codeMantra

Publisher's Note
The publisher accepts responsibility for any inconsistencies that may have arisen during the conversion of this book from journal articles to book chapters, namely the inclusion of journal terminology.

Disclaimer
Every effort has been made to contact copyright holders for their permission to reprint material in this book. The publishers would be grateful to hear from any copyright holder who is not here acknowledged and will undertake to rectify any errors or omissions in future editions of this book.

Contents

Citation Information

The following chapters were originally published in different issues of the *Water International* and *International Journal of Water Resources Development*. When citing this material, please use the original citations and page numbering for each article, as follows:

Chapter 1
Putting water security to work: addressing global challenges
Chad Staddon and Christopher A. Scott
Water International, volume 43, issue 8 (2018) pp. 1017–1025

Chapter 2
Is water security just? Concepts, tools and missing links
Sarah Wade
Water International, volume 43, issue 8 (2018) pp. 1026–1039

Chapter 3
Challenges in meeting water security and resilience
Vijay P. Singh
Water International, volume 42, issue 4 (2017) pp. 349–359

Chapter 4
The cultural dimensions of household water security: the case of Kathmandu's stone spout systems
Olivia Molden, Nicholas Griffin and Katie Meehan
Water International, volume 41, issue 7 (2016) pp. 982–997

Chapter 5
Contested access: improving water security through benefit sharing
Bimo A. Nkhata
Water International, volume 43, issue 8 (2018) pp. 1040–1054

Chapter 6
Why doesn't every family practice rainwater harvesting? Factors that affect the decision to adopt rainwater harvesting as a household water security strategy in central Uganda
Chad Staddon, Josh Rogers, Calum Warriner, Sarah Ward and Wayne Powell
Water International, volume 43, issue 8 (2018) pp. 1114–1135

Chapter 7

Water security and the pursuit of food, energy, and earth systems resilience
Christopher A. Scott, Tamee R. Albrecht, Rafael Routson De Grenade, Adriana Zuniga-Teran, Robert G. Varady and Bhuwan Thapa
Water International, volume 43, issue 8 (2018) pp. 1055–1074

Chapter 8

Unraveling transboundary water security in the arid Americas
Tamee R. Albrecht, Robert G. Varady, Adriana A. Zuniga-Teran, Andrea K. Gerlak, Rafael Routson De Grenade, América Lutz-Ley, Facundo Martín, Sharon B. Megdal, Francisco Meza, Diego Ocampo Melgar, Nicolás Pineda, Facundo Rojas, Rossi Taboada and Bram Willems
Water International, volume 43, issue 8 (2018) pp. 1075–1113

Chapter 9

De-nationalization and de-securitization of transboundary water resources: the Israeli–Palestinian case
David B. Brooks and Julie Trottier
International Journal of Water Resources Development, volume 30, issue 2 (2014) pp. 211–223

Contributors

Tamee R. Albrecht School of Geography and Development & Udall Center for Studies in Public Policy, University of Arizona, Tuscon, USA.

David B. Brooks International Institute for Sustainable Development, Ottawa, Canada.

Rafael Routson De Grenade Udall Center for Studies in Public Policy, University of Arizona, Tucson, USA.

Andrea K. Gerlak School of Geography and Development & Udall Center for Studies in Public Policy, University of Arizona, Tucson, USA.

Nicholas Griffin Wake Forest University School of Law, Winston Salem, USA.

América Lutz-Ley Center for Studies on Development, Colegio de Sonora, Hermosillo, Mexico.

Facundo Martín Department of Geography, Universidad Nacional de Cuyo & Consejo Nacional de Investigaciones Científicas y Técnicas, CONICET, Mendoza, Argentina.

Katie Meehan Department of Geography, University of Oregon, Eugene, USA.

Sharon B. Megdal Water Resources Research Center & Department of Soil, Water and Environmental Science, University of Arizona, Tucson, USA.

Diego Ocampo Melgar Oceans and Atmosphere Department, Commonwealth Scientific and Industrial Research Organisation, Santiago, Chile.

Francisco Meza Facultad de Agronomía e Ingeniería Forestal, Pontificia Universidad Católica de Chile, Santiago, Chile.

Olivia Molden Department of Geography, University of Oregon, Eugene, USA.

Bimo A. Nkhata Monash South Africa, Roodepoort, South Africa.

Nicolás Pineda Centro de Estudios en Gobierno y Asuntos Públicos, Colegio de Sonora, Hermosillo, Mexico.

Wayne Powell Department of Geography and Environmental Management, University of the West of England, Bristol, UK.

Josh Rogers Department of Geography and Environmental Management, University of the West of England, Bristol, UK.

Facundo Rojas Instituto Argentino de Nivología, Glaciología y Ciencias Ambientales (IANIGLA-CONICET) & Departamento de Geografía, Facultad de Filosofíay Letras, Universidad Nacional de Cuyo, Mendoza, Argentina.

Christopher A. Scott School of Geography and Development & Udall Center for Studies in Public Policy, University of Arizona, Tuscon, USA.

Vijay P. Singh Department of Biological and Agricultural Engineering and Zachry Department of Civil Engineering, Texas, USA. A&M University, College Station, USA.

Chad Staddon Department of Geography and Environmental Management, University of the West of England, Bristol, UK.

Rossi Taboada Programa Agua-Andes, Centro de Competencias de Agua, Lima, Peru.

Bhuwan Thapa School of Geography and Development & Udall Center for Studies in Public Policy, University of Arizona, Tuscon, USA. Indiana University, Indianapolis, USA.

Julie Trottier Progamme de l'Eau, Centre National de la Recherche Scientifique and Centre de Recherche Francais a` Jerusalem, Jerusalem, Israel.

Robert G. Varady Udall Center for Studies in Public Policy, University of Arizona, Tucson, USA.

Sarah Wade is the global Sector Lead at the Alliance for Water Stewardship (AWS). She studied Water Security and International Development at the University of East Anglia and went on to work at WWF-UK prior to joining AWS in 2018.

Sarah Ward is a CIWEM-chartered PhD, whose research, teaching and practice focuses on water, environment, communities and resilience. Her work cuts across systems, scales and disciplines, using systems thinking, technology assessment and sustainability transitions and applying mixed methods and participatory approaches. Sarah specialises in making, visualising and advocating for recognition of the complex connections between water, energy, carbon, infrastructure, sustainability and society in the context of climatic and environmental change. Sarah is also an advocate for inclusivity, diversity and equality in the water and environment sector, recently co-founding CIWEM's WEM Pride network, which aims to provide a voice for and support to LGBTQ+ individuals in WEM spaces.

Calum Warriner Department of Geography and Environmental Management, University of the West of England, Bristol, UK.

Bram Willems Programa Agua-Andes, Centro de Competencias de Agua, Lima, Peru.

Adriana A. Zuniga-Teran Udall Center for Studies in Public Policy & School of Landscape Architecture and Planning, University of Arizona, Tucson, USA.

Introduction: Rethinking Water Security in a Globalised World: Steps to Meeting SDG6

Chad Staddon and Christopher A. Scott

1 Introduction: why talk about water security?

Notwithstanding significant programmatic progress made with the Sustainable Development Goals, and in particular SDG 6 "clean water and sanitation for all", over the last decade, water "security" has become a key optic for thinking about how we manage water, supplanting past approaches such as Integrated Water Resources Management. The water security reframing has offered benefits including clear recognition of the link between humans and the environment, the human right to water and a renewed sense of urgency in equitably and safely providing water while mitigating water-related disaster. This has also posed challenges including the tendency in some quarters to interpret "security" in terms of geopolitical and economic "securitisation", the latter linked closely with the neoliberal agenda of subjecting management of water to the dictates of the market. There is also emotional baggage associated with the notion of water security. After all, as rhetoric, "sustainability" somehow seems less immediate, less personal, less potentially threatening than "security". And security can focus the (political) mind in at least two quite immediate ways: in terms of the prospect of improving levels of water security for a given community, but also in terms of the risk of going in the other direction, towards water insecurity.

According to a standard definition used by UNESCO (2008):

> Water security involves protection of vulnerable water systems, protection against water related hazards such as floods and droughts, sustainable development of water resources and safeguarding access to water functions and services.

Thankfully this definition and others like it (see Gerlak et al., 2018; Staddon & James, 2013 for broader reviews) emphasise the importance of the nexus between people and the environment for achieving any enduring sense of water security. In an age of climate emergency, this seems a positive step. Ultimately this framing leads also inexorably towards political resolutions and political contestations. Note that the UNESCO definition rests on an acceptance that something called "water" is somehow under threat and that the designating actors, in other words, us humans, are central to addressing that threat. As is so often the case in the history of our relations with the natural world, the (non-human) hydrosphere is largely passive, whilst we humans are active, endowed with agency, and - hopefully - responsibility.

Though it has a longer and more complex genealogy, the water security concept really took centre stage after the 2011 Bonn Conference. However, missing from early framings of water security were notions of distributional justice (explored by Wade in this volume), cultural resonance (explored by Molden et al.), intersectionalities (Thapa in this volume), and scale (Albrecht et al. in this volume). In this brief introduction to the present expanded *Special Issue as Book*, we discuss some of the key theoretical intersections raised by the chapters that follow, point to key emerging areas of water security research, and suggest where water security research may go in the future.

2 Genealogy of the "Water Security" Concept

As Staddon and James note (2013), the first explicit invocations of the water security concept were associated with publications about the relations between water scarcity and political conflict in the Middle East appearing in the early 1990s (e.g. Anderson, 1991; Shuval, 1992). The potential for scarce water resources to exacerbate regional conflict in the Middle East had been generally recognised[1] long before the term water security was first coined in 1991 by foreign policy analyst Joyce Starr in an article explicitly linking "water" and "security" to the longer-term prognosis for stability in the Persian Gulf region, then still in the throes of the First Gulf War. After reciting demographic statistics to underpin the (neo-Malthusian) claim that regional population will soon outstrip available water resources, Starr then suggests a congruity between water security and military security: "Water security will soon rank with military security in the war rooms of defense ministries" (Starr, 1991, p. 19). Others, especially from the political science community (e.g. Bulloch & Darwish, 1993; Klare, 2001; Ohlsson, 1995) have developed further the geopolitical dimensions of the water security concept in ways that place water at the centre of thinking about regional stability.

The emergence of water security as a dominant framing concept was not however solely the product of events in the Middle East. At least three other factors have helped to both deepen and widen the hold of the "water-security nexus" on the policy, scholarly and popular imaginations. The first is the simple fact that by the early 1990s other world regions were experiencing their own versions of water insecurity. For example, in the Mekong Basin, where the world's 10th longest river is shared by six nations all vying for access to the water resources necessary for development, tensions have been growing over the claims of upstream riparians to ever-increasing water shares. Two of these riparian nations, Myanmar (Burma) and China, have proven less than fully cooperative with attempts by the other four riparians to jointly manage the river's resources (Jacobs, 2002). And there are also sub or intra-regional examples of pending water insecurity, such as the Colorado River Basin, where the original 1922 inter-state water compact turned out to allocate water volumes based on unrealistically high estimates of annual river flow. This fact, together with new science on environmentally necessary river flows have forced the seven American states and Mexico (at the mouth of the Colorado River at the Sea of Cortez) to find new ways of optimally allocating reduced water volumes from the river and has been centrepiece of relations between the US and Mexico (Wilder et al., 2020; Wildman & Forde, 2012). Here, we note the imperative to expand research and applications of water security for rivers and aquifers that are transboundary and trans-jurisdictional.

The second factor which has helped propel the development of the "water security" concept is the (too slow) realisation of the seriousness of climate change impacts on long-term water availability in many parts of the world. This factor has already been alluded to above, heightening pre-existing tensions around shared water resources, whether at the international or subnational levels. The IPPC 4th Assessment, released in late 2007, suggested that we have already seen a 0.74 °C increase in global average temperature since 1900 (IPCC, 2007). By the time of the 5th IPCC Assessment, published in 2014, that estimate had been doubled to more than 1.5°C (IPCC, 2014). While there is much debate about the precise nature of the interactions between key factors such as global CO_2 concentrations, the Jet Stream, and the Atlantic Thermohaline Circulation, there is no longer any doubt that we live in an era of Anthropogenic warming with profound implications for the hydrosphere and all life on earth that depends on it.

Climate change also means that we are likely to experience an increase in the intensity, severity and frequency of extreme weather events such as droughts, storms and floods. The latter raises the need for future water security research on the rising frequency with which water's destructive impacts are felt. And flooding's Janus-faced twin, drought, is also intensifying. Regions at risk of drought are likely to become larger, for example, taking in much of southern Europe, the Mediterranean Basin and even swathes of central Europe (EEA, 2017; Lloyd-Hughes & Saunders, 2002). Differences in water availability between regions will become increasingly pronounced. Areas that are already relatively dry, such as the Mediterranean basin and parts of Southern Africa, South America and North America, are likely to experience further decreases in water availability. For example, several (but not all) climate models predict up to a 30% decrease in annual runoff in these regions alongside a 2°C global temperature rise. In contrast, parts of Northern Europe and Russia are likely to experience increases in water availability (runoff), for example, a 10–20% increase for a 2°C temperature rise. In all cases however existing drought conditions will intensify, particularly those regions affected by winter droughts (Central Eastern Europe, Western Russia, North-western Europe including Norway). What's more, climate change awareness has come concomitantly with increased awareness of the substantial water needs of the natural environment. Thus, even climate-conscious water managers are finding themselves hard-pressed to accommodate all demands with a shrinking resource.

The third factor which has helped push the concept of water security ever up the agenda has been the realisation that water is central to so many of life's other necessities. Thus, we now have the "water-energy nexus", the "water and food security nexus" (Anon., 2011; Hadley, 2009; Martinez-Cortina, Garrido, & Lopez-Gunn, 2010; Shrivastava, 2003), "tourism and water security" (Gossling et al., 2012), "water quality security" (Rose, 2002) and even the security of water services infrastructure itself (Skolicki, 2008). Briefly put, the "water-energy nexus" creates an indissoluble link between water security and energy security inasmuch as energy production requires water inputs and water services production requires energy inputs. Similarly, water is required for food production and food is (of course) required for the reproduction of the labour power that manages water services. The World Economic Forum meetings at Davos in winter 2011 identified an almost mind-boggling proliferation of "nexuses" of which water was one half of the binary including "water-food", "water-energy", "water-trade" and "water-health" (WEF, 2011). Academic interest in the water-energy-food (WEF) nexus (cf. Hoff, 2011) has rapidly increased over the last decade (e.g. Albrecht, Crootof, & Scott, 2018; Bleischwitz, Hoff, Spataru, van der Voet, & VanDeveer, 2018; Dermody et al., 2018; Endo et al., 2015; Scott et al., this volume), usually with the aim of better understanding and quantifying trade-offs and synergies in these complex, multi-scale, interactional systems.

Based on the above it is clear that several "roads" – political, climatic, social and scientific – have led to the current popularity, and therefore discursive power, of the water security concept. This set of branching paths, as it were, set the stage for new understandings of water security, set in broader social and political context. Paraphrasing the environmental philosopher Tim Luke, we suggest that:

> Encircled by grids of ecological alarm, [water security] discourse tells us that today's allegedly unsustainable environments need to be disassembled, recombined and subjected to the disciplinary designs of expert management (Luke, 1999, p. 142).

These "grids of ecological alarm" are simultaneously political, social, technological and natural involving a dangerously degenerative set of interconnected dynamics that are now working together to undermine the long-term prospects for well-being and security everywhere on earth. Luke warns us to be mindful of the "disciplinary designs of expert management". As the historian Michel Foucault put it, we need to carefully unpick the complex web of:

> ...people in their relations, their links, their imbrication with those other things which are wealth, resources, means of subsistence, the territory with its specific qualities, climate, irrigation, fertility, etc.; [people] in their relation to customs, habits, ways of acting and thinking, etc.; lastly [people] in their relation to that other kind of thing, accidents and misfortunes such as famine, epidemics, death, etc. (Foucault, 1991, p. 93).

We suggest that the ascendance of the water security concept marks not just a fashion, but rather an epistemological paradigm shift. As is always the case, such moments make negative as well as positive changes possible. This is why we are so concerned about the potential for the water security concept to be co-opted to the demands of the revanchist populisms recently seen in the US (Trump), UK (Johnson), Turkey (Erdogan), China (Xi) and elsewhere. It is critically important that populist narratives of water security are seen for what they are: as attempts to privatise certain use values of water for the sole disposition of the new elites that have emerged in the first decades of the twenty-first century. As Foucault (in Dreyfus & Rabinow, 1984, p. 5) tells us:

> the real political task in a society such as ours is to criticize the working of institutions which appear to be both neutral and independent; to criticize them in such a manner that the political violence that has always exercised itself obscurely through them will be unmasked, so that one can fight them.

We see special collections such as this one as critical discursive interventions with the power to show that the seemingly inevitable state of things can and is being challenged by framings of water security deeply rooted in appreciation of water's socio-natural and emancipatory realities.

3 Putting Water Security to Work!

This volume brings together authors primarily (but not only) from the social sciences involved in critically rethinking the water security concept. While most of the articles were presented at the 2015 IWRA World Water Congress in Edinburgh, this edited collection also includes new additions (including this revised introduction), chosen because they help illuminate areas not covered by the original collection. The 10 chapters in this volume represent the collective enterprise of over two dozen authors examining key themes including how water security articulates with locality and culture, how it operates across spatial scales and its intersections with concepts of justice, nexus, diplomacy and benefit-sharing.

In his 2017 Ven Te Chow Memorial lecture to the IWRA World Water Congress in Cancún, Mexico, Vijay Singh provides a synoptic overview of high-level water security challenges including water availability for human health, energy, food and other competing uses in different parts of the world. Addressing these challenges requires the development of a new "…water, food and energy security system, partnerships among academia, the government sector, NGOs and the private sector (farmers). The educational system that trains the trainers and teaches the teachers must be emphasized." If practitioners and scholars broadly agree that water security is a multi-disciplinary challenge, then there is a strong role for organisations like IWRA to help develop and deliver the required integrative capacity.

In her contribution to this Special Issue, Sarah Wade argues strongly for framing water security as a "justice" issue, very much in procedural terms that focus on free and open deliberation that can "…draw in the perspective[s] of stakeholders across a variety of levels and consider the power dynamics influencing their engagement with water" (Wade, this volume). For Nkhata, water justice and security are not abstractions but are directly linked to the manner in which benefits deriving from the use of scarce water resources are *shared* amongst otherwise competing beneficiaries. His case study of contestation amongst tour boat operators on the Lower Zambezi National Park suggests that Ostrom-type allocative systems can result in the generation of more benefits that are also better distributed amongst competing operators.

Staddon et al. highlight the most important factors in helping to create more equitable access to opportunities for expanding local water supply systems. Their examination of factors associated with uptake of rainwater harvesting by households suggests that locally-embedded intermediary organisations are of paramount importance. Their social, as well as economic and political capital, seems to be the key determinant, over and above finance availability, family size, level of education, and other commonly noted determinants of uptake. Traditional water systems, too, can play an important role in addressing contemporary water challenges, as Molden et al.'s paper on Nepali "water spouts" illustrates. More than just celebrating their physical performance, Molden et al. note that Nepali water spouts (and by extension, Iranian qanats, Latin American acequias, Rajasthani Taanka, etc.) are linked to systems of religious, social and political self-organisation that are often challenged by central state authorities unable to see their enduring value.

Implicit in the above chapters is the reality that water is never only one thing. For some stakeholders, water is about addressing immediate household requirements, whilst for others, it could be about ensuring sufficiency of water for food or energy production. Indeed, the central importance of the intersectionalities between water (as wet stuff, as hydraulic head, as a carrier of agricultural sediments, etc.) and other sectors usually considered separately (e.g. power generation, food production) has spawned its own intellectual growth industry – the "water-energy-food nexus". Scott et al.'s chapter offers a comprehensive review of the emergence of water security following food and energy security concepts – now intertwined in the WEF nexus – in contemporary resource scholarship and advocates "soft path" approaches to resolving competition between competing stakeholders:

> nexus approaches should be process-oriented and address dynamic interplays in research and policy. These approaches should also be flexible enough to account for extreme environmental and social conditions, which will become more commonplace with global change. (Scott et al., this volume)

Ultimately, we need to abandon narrow specialisms in favour of more broadly trained water (and energy, food, etc.) management professionals so that intersectionalities rather than monofunctions come to dominate science, policy and science-policy discourse.

What emerges from the first six chapters in this collection is a clear call for more attention to the multiple intersectionalities linked to effective hydrogovernmentality under competing challenges of climate change and social justice. To put it another way, water governance is not just a matter for central authorities acting "according to the science", as is so often said these days (usually to stifle further debate!). Albrecht et al.'s chapter on transboundary water governance in the arid Americas shows that whilst clear legal rights and responsibilities are important, it is equally important that water governance is practised by subnational stakeholders who see well beyond international boundaries. Their unbundling of transboundary water security challenges into complex intersecting "water security attributes" offers a set of conceptual templates for thinking about transboundary water security challenges worldwide. Brook and Trottier draw on case studies of hydropolitics in a different part of the world, the Middle East, to further develop the key point that transboundary water management is not inevitably a *casus belli*. More to the point, their chapter makes clear that successful water-sharing across boundaries is more likely where allocation rules (e.g. treaties) are not prescriptive, but procedural and where there is room for informal working parties and interactions outside the purview of political representatives.

4 Whither Water Security?

As with most conceptual and applied frameworks, water security is experiencing rapid transformations, driven by the diversity of scholars and practitioners who employ it for analytical value and who otherwise "put it to work." This raises the need to adapt water security conceptually and to continually test its ability to shed better light on current and future challenges. In this brief introductory paper for the SPIB, we have identified three main areas we see as emerging needs and opportunities for water security research: (1) refined understanding of transboundary and trans-jurisdictional applications of water security, (2) security from the destructive phases of the water cycle heightened by climate change, and (3) broader societal framings of water security beyond resource linkages to energy and food security. We address here each of these three dimensions of the future of water security research.

Water security has been addressed both as a local and a regional challenge in resource, administrative and political terms. The transboundary aspects reviewed by Albrecht et al., for example, show how multiple local water securities, as it were, combine in novel ways to frame broader transboundary security challenges. Yet, additional, intersecting securities – for example, of water bodies in ecological terms (rivers, lakes, aquifers themselves) or of contrasting notions of water security among multiple transboundary interests (equity vs. efficiency) – are not fully addressed or reconciled in current assessments. Future efforts could well gain insights by considering broader societal and environmental dimensions of transboundary and other trans-jurisdictional water securities, a point we consider in the third area of needed research, below.

The second important area of research on water security is the relative paucity of applications for destructive, high-risk and high-impact contingencies, particularly from flooding, severe weather and extreme events. Grey and Sadoff (2007) had observed this, and more recent bibliometric analyses (Gerlak et al., 2018) have demonstrated the lack of, and gaps in, understanding of water insecurity resulting from acute and chronic flooding. As we noted above, climate change will force the hand of water managers and more importantly, the public and society at large, to plan for the security from flooding of humans (including people, communities, settlements and infrastructure) and ecosystems (particularly threatened species and habitats not adapted to flooding). We see this as an inevitability of resilience understandings of recovery and regrouping after disturbance and disaster. Here, water security must gain closer linkages with disaster risk and resilience framings.

The third area in which water security research and applications are likely to evolve, and are much needed, are the broader social and political implications of the concept, that is, the renewed and heightened relevance of water security in humanistic (arts and culture), human rights, and political discourse. Related to our previous two points, especially the first on the trans-jurisdictional imperative, broader framings of water security must be understood in process terms not simply in static (or status) terms of metrics of attainment. In other words, conceptual work and applications are needed for water security in dynamic planning contexts that may include leveraging of the approach by interests beyond academia and formal decision-makers. Akin to the way in which energy security and food security, for example, now have broad public understanding and acceptance, research and dissemination are needed that will poise water security as a popular, electoral and political device. In sum, we are optimistic for the broadening and deepening of water security in the coming years.

Finally, we feel we must acknowledge that we are launching this collection into a radically different world than that of the 2017 IWRA conference where it originated.

The campaign by Black Lives Matter to force all societies around the world to confront and challenge entrenched inequalities and the 2020–21 Coronavirus pandemic both have significant implications for water security scholarship. BLM requires us to go beyond acknowledgement and fundamentally challenge resource inequalities. No longer is it sufficient to just point to systemic forces behind unequal water access. We must also recognise that the structures of consumer power in a neoliberal world are as impactful as the decisions of sinister state agents in depriving parts of the world of the water resources they need to develop and prosper. The urgencies of moving from critique to ethics and then to action are greatly magnified during the current pandemic, a pandemic that should teach all Earth's citizens that there can be no water security for some unless there is water security for all.

Acknowledgments

We acknowledge support from the International Water Security Network, funded by Lloyd's Register Foundation, a charitable foundation helping to protect life and property by supporting engineering-related education, public engagement, and the application of research. Additional support was provided by the Inter-American Institute for Global Change Research (IAI) CRN3056 supported by the US National Science Foundation (Grant GEO-1128040).

Note

1. See for example Allan (1983).

References

Albrecht, T. R., Crootof, A., & Scott, C. A. (2018). The water-energy-food nexus: A systematic review of methods for nexus assessment. *Environmental Research Letters, 13*, 043002. doi:10.1088/1748-9326/aaa9c6

Allan, J. A. (1983). Natural resources as national fantasies. *Geoforum, 14*(3), 243–247.

Anderson, E. W. (1991). The Middle East and hydropolitics. *World Energy Council Journal*, December, 35–38.

Anon. (2011). *Water security: The water-food-energy-climate nexus.* [online].

Bleischwitz, R., Hoff, H., Spataru, C., van der Voet, E., & VanDeveer, S. D. (Ed.). (2018). *Routledge handbook of the resource nexus.* London and New York: Routledge, 517pp.

Bulloch, J., & Darwish, A. (1993). *Water wars: Coming conflicts in the Middle East.* London: Victor Gollancz.

Dermody, B. J., Sivapalan, M., Stehfest, E., van Vuuren, D. P., Wassen, M. J., Bierkens, M. F. P., & Dekker, S. C. (2018). A framework for modelling the complexities of food and water security under globalisation. *Earth System Dynamics, 9*, 103–118. doi:10.5194/esd-9-103-2018

Dreyfus, H. L., & Rabinow, P. (Eds.). (1984). *Michel Foucault: Beyond structuralism and hermeneutics* (2nd ed.). Chicago: The University of Chicago Press.

EEA. (2017). Climate change, impacts and vulnerability in Europe 201: An indicator-based report.

Endo, A., Burnett, K., Orencio, P. M., Kumazawa, T., Wada, C. A., Ishii, A., . . . Taniguchi, M. (2015). Methods of the water-energy-food nexus. *Water, 7*, 5806–5830. doi:10.3390/w7105806

Foucault, M. (1991). Governmentality. In G. Burchell, C. Gordon, & P. Miller (Eds.), *The Foucault effect: Studies in governmentality* (pp. 87–104). Chicago: The University of Chicago Press.

Gerlak, A. K., House-Peters, L., Varady, R., Albrecht, T., Zuniga Teran, A., de Grenade, R., . . . Scott, C. A. (2018). Water security: A review of place-based research. *Environmental Science & Policy, 82*, 79–89. doi:10.1016/j.envsci.2018.01.009

Grey, D., & Sadoff, C. W. (2007). Sink or swim? Water security for growth and development. *Water Policy, 9*(6), 545–571, 27p.

Gossling, S., Peeters, P., Hall, C. M., Ceron, J. P., Dubois, G., Lehmann, L. V., & Scott, D. (2012). Tourism and water use: Supply, demand, and security. An international review. *Tourism Management* [online], *33*(1), 1–15.

Hadley, C. (2009). Experience-based measures of food and water security: Biocultural approaches to grounded measures of insecurity. *Human Organization* [online].

IPCC. (2007). Climate Change 2007: Synthesis Report. Contribution of Working Groups I, II and III to the Fourth Assessment Report of the Intergovernmental Panel on Climate Change [Core Writing Team, Pachauri, R.K and Reisinger, A. (Eds.)]. IPCC, Geneva, Switzerland, 104pp.

IPCC. (2014). Climate Change 2014: Synthesis Report. Contribution of Working Groups I, II and III to the Fifth Assessment Report of the Intergovernmental Panel on Climate Change [Core Writing Team, R.K. Pachauri and L.A. Meyer (Eds.)]. IPCC, Geneva, Switzerland, 151pp.

Jacobs, J. W. (2002). The Mekong River Commission: Transboundary water resources planning and regional security. *Geographical Journal, 168*, 354–364. doi:10.1111/j.0016-7398.2002.00061.x

Martinez-Cortina, L., Garrido, A., & Lopez-Gunn, E. (2010). Re-thinking water and food security: Fourth Botin Foundation water workshop [online].

Ohlsson, L. (1995). *Hydropolitics: Conflicts over water as a development constraint*. Dhaka, Bangladesh: The University Press Ltd.

Rose, J. B. (2002). Water quality security. *Environmental Science and Technology* [online].

Scott, C. A., Albrecht, T., de Grenade, R., Zuniga-Teran, A., Varady, R. G., & Thapa, B. (2018). Water security and the pursuit of food, energy, and earth systems resilience. *Water International, 43*, 1055–1074. doi:10.1080/02508060.2018.1534564

Shrivastava, G. S. (2003). Water resources and food security: A Caribbean case study. *Proceedings of the Institution of Civil Engineers: Water and Maritime Engineering* [online].

Skolicki, Z. (2008). Co-evolution of terrorist and security scenarios for water distribution systems. *Advances in Engineering Software* [online], *39* (10), 801–811.

Starr, J. R. (1991). Water wars. *Foreign Policy, 82* (Spring), 17–36.

Wilder, M. O., Varady, R. G., Gerlak, A. K., Mumme, S. P., Flessa, K. W., Zuniga-Teran, A. A., . . . Megdal, S. B. (2020). Hydrodiplomacy and adaptive governance at the US-Mexico border: 75 years of tradition and innovation in transboundary water management. *Environmental Science & Policy, 112*, 189–202.

Wildman, R. A., Jr. & Forde, N. A. (2012). Management of water shortage in the Colorado River basin: Evaluating current policy and the viability of interstate water trading1. *JAWRA Journal of the American Water Resources Association, 48*, 411–422. doi:10.1111/j.1752-1688.2012.00665.x

World Economic Forum (2011). *Water security: The water-food-energy-climate nexus*. Earthscan Press.

Putting water security to work: addressing global challenges

Chad Staddon and Christopher A. Scott

Introduction

Over the last decade, the concept of water security has emerged from its origins in a niche of studies of international security and hydropolitics to become a common currency of researchers and policy makers. Indeed, in some quarters it seems even to be supplanting the dominant position hitherto occupied by 'sustainable water management', as well as the Integrated Water Resources Management framework. A simple trawl of international water-themed conferences revealed that, as early as 2012, seven used the term 'security', while only four used the term 'sustainability'. Sustainability is, it seems, out of fashion, while security is all the rage. Moreover, problems as otherwise diverse as US–Mexican border relations, undocumented migration into Europe, malnutrition in the global South and growing energy demand have been increasingly reframed as problems of environmental security. Since the 1990s Canadian and American security services have often asserted that environmental change can be a territorial security threat, particularly with regard to growing conflict over scarce resources ('resource wars') and in-migration (cf. Homer-Dixon, 1994; Klare, 2001). Global institutions such as FAO, UNEP, UNESCO, the World Water Council, Global Water Partnership, World Bank and World Economic Forum have enthusiastically seized on the concept of water security (Gerlak et al., 2018) – as have regional groupings such as the Gulf Coordination Council and the Asian Development Bank – establishing it as a central organizing principle for water management in the twenty-first century. So, sustainable water, with its Brundtland-era associations with intergenerational equity and the triple bottom line, seems to have shifted to backstage, and water security is now very much 'in'.

This special issue of *Water International* is predicated on the contention that water security is a concept that upon careful examination reflects a conflict of world views, and with it, of political approaches. Since it appears to be a paradigm for water management in the twenty-first century it is crucially important that users are clear about what it does, and does not, signify. In particular we argue that it is important to better understand the way the concept may be deployed to replace the more democratic water governance implied by 'sustainable development' with a state-dominated and 'territorialist' form of environmental governmentality. This brief essay introduces the concept of water security in terms of its history and the main themes associated with it, and offers a critical multidisciplinary reconceptualization which could provide needed common ground within a burgeoning discourse already showing signs of friction and rupture. Further analysis and critical reflection on the emergence of water security along with energy and food security are presented by Gerlak et al. (2018), Varady, Albrecht, Gerlak, Zuniga, and Staddon (in press), Staddon and James (2014), and Scott et al. (this issue).

A critical perspective on water security

The securitization of environmental discourse is part of a decades-long process that is not wholly coincident with the sustainable development movement and indeed seeks to reposition the state and civil society as key actors in environmental management. Though there are antecedents, the debate on the securitization of environmental issues emerged particularly in the 1990s, along with the publication of works by Norman Myers (1994) and Hillel Shuval (1992). As Liverman (2009) points out, there is a danger, and a reality in some cases, that discourses of environmental (or water) security may take us back to the days when it was often asserted that climate dictates culture and character. Speaking on behalf of an entire generation of imperialists in the early twentieth century, the geographer Victor Cousins put it like this (quoted in Febvre, 1932, p. 10):

> Give me the [physical] map of a country … and I pledge myself to tell you, a priori, … what part that country will play in history, not by accident, but of necessity; not at one epoch, but in all epochs.

Unsurprisingly, theorists of this school tended to give the most flattering characterizations to the European and North American nations of the middle and upper latitudes. Of course much work has since been done to explore the complex relationships between physical and social spaces, but it is clear that in some corners a simple determinism endures. After all, judging by the popularity of recent work linking social change to climate change (e.g. Afolayan & Adelekan, 1999; Diamond, 2005; Ezra, 2001; Homer-Dixon, 1994), often resting on a Social Darwinist foundation,[1] we feel compelled to caution that glibly made links between climate, conflict and migration are not empirically justified and raise concerns about the securitization of climate discourse. This securitization, we argue, can operate in at least two ways:

(1) (Overt) political securitization of water issues – where a state, or prominent actors within it, found policy towards an extra-territorial 'other' on the view that environmental change (short or long term) is to blame. Such views are likely to see resources as things to be hoarded and/or violently acquired, and environmental politics as a zero-sum game. This view is most likely to see links between environmental change and conflict, including armed conflict.
(2) Economic securitization of water issues – where a state, or prominent actors within it, may not adhere to point 1 above, but view correct policy as policy that

uses market mechanisms to allocate increasingly scarce resources, and indeed transforms environmental assets into financial ones to facilitate this process. This is essentially Washington Consensus ideology[2] and proclaimed loudest by those national actors who feel they have the economic power to 'win' this game.

Of course, these realist and neo-realist international relations discourses can overlap within a single larger discourse formation, with the resultant danger that more progressive theorizations of water security are squeezed out.

Our analytical perspective draws heavily on critical political ecology, defined here, following Tim Forsyth (2003, p. 20) and others, as the 'development of an analytical approach that is biophysically grounded yet conscious of social and political constructions' of such key terms as 'environment', 'nature' and 'natural resources'. Beyond incorporating the marginalized and disenfranchized, the postulation of a 'critical political ecology' also signals a deeper concern with the ways in which meaningful discourses *about* the hydrosocial world are produced, which subsequently result in material practices of management and exploitation *of* that same hydrosocial world. A higher-level summation, to which we will return in the concluding section of this introduction, asks us to reframe water security in terms of *security of what and for whom.* From our perspective, 'water security' shares with its 'sustainable water' predecessor certain problems of political ontology with which progressive scholars and activists have long grappled. Challenging these deficiencies is not an option – progressive scholars must attempt to take the water security concept away from realist and neorealist geopoliticians, both through reframing the ontological problems it is meant to address, and through epistemologically resituating it with respect to what Ranciere and Swyngedouw have called the 'postpolitical'. In other words, we seek to challenge politically regressive framings by proposing more emancipatory and democratic ones. We are not the only ones exploring this terrain; Beck, Foucault, Latour and others have travelled here too. Chantal Mouffe (2005, pp. 40–41) summarizes post-democracy as follows:

> What is needed is the creation of forums where a consensus could be built between the experts, the politicians, the industrialists and citizens on ways of establishing possible forms of co-operation among them. This would require the transformation of expert systems into democratic public spheres.

This reframing is of critical importance for thinking about environmental security and water security because it strikes at the very heart of what is at stake: the politics of the changing relationship between society and the natural world. It is more than a decade since Lucy Jarosz (2004), citing Marc Auge, asked the key ethical question: 'How do we retain a sense of the other and a sense for the other in terms of environments and the societies living within them? How do we understand ourselves and others in critical and compassionate ways?'

This is difficult terrain, not least because different participants in this burgeoning discourse have tended to mobilize the concept differently. Political scientists and governments have often tended to prioritize the securitization dimension of water security, seeing it in terms of the relation between water and political and/or military interactions between states, pre-eminently in the contemporary Middle East. These studies and policy orientations move the concept close to the idea of hydropolitics or even, as Schulz (1995) put it, the 'hydropolitical security complex' (see also Klare, 2001; Ohlsson, 1995). Others have conceived of water security in terms of security *from* water-related risks and hazards:

> Achieving basic water security, both harnessing the productive potential of water and limiting its destructive impact, has always been a societal priority. To capture this duality, water security is defined here as the availability of an acceptable quantity and quality of water for health, livelihoods, ecosystems and production, coupled with an acceptable level of water-related risks to people, environments and economies. (Grey and Sadoff, 2007)

But who is to decide where the line between acceptable and unacceptable water-related risks can be drawn? Is it acceptable that the risks of drought or flood, however objectively large or small, are disproportionately borne by the most marginalized and politically weak parts of society? Whereas the discourse of sustainable water is often talked about in terms of the triple bottom line, the water security discourse more readily overlooks issues of social impact. Desperately needed are democratic mechanisms for both defining the boundaries between acceptable and unacceptable outcomes and giving the resulting laws and policies social legitimacy.

Consider the role that normalizing science has played in this transition from ecological modernism, to sustainable development, to environmental (and water) security. Scientific knowledge production tacks back and forth between the evolving understandings of individual researchers, teams and institutes and the demands and constraints imposed by institutions, including funding institutions and accreditation bodies, to characterize the water security issue in particular ways. The endeavours of mapping, measuring, calculating, and providing models make the multiple identities of water and the environment quantifiable, manageable and controllable. Water then becomes a resource knowable only through the respective kinds of expertise developed by ecologists, biologists, geologists, engineers and agronomists, who tend to triangulate onto the same discourse of crisis and response (Cook and Bakker, 2012). Repetition and the piling up of opinion certainly trends in apparently preordained directions. Alternative knowledges, such as those not framed within Euclidean space, find it increasingly difficult to gain traction: for example traditional or non-Christian knowledges of water's ontology. In other words, 'what was unknown has become fully knowable; what was mysterious is now readily imaginable; *and the whole has become eminently governable*' (Peace, 2002, pp. 536–37). This governability of water easily ties itself in with concerns with carrying capacity and limits to resources, or as Luke (1999, p. 142) has put it: 'Encircled by grids of ecological alarm, [climate change discourse] tells us that today's allegedly unsustainable environments need to be disassembled, recombined and subjected to the disciplinary designs of expert management.'

Here we would point to the all-too-naturalized tendency of most policy makers and scholars (ourselves included) to found their work on the assumed triple threat of climate-change-induced water scarcity, population rise and declining ecosystem services values. And governmentality does not need to be total or smooth, as a study of water politics in post-communist Bulgaria revealed: 'The Djerman-Skakavitsa [water diversion] controversy reveals ragged, unstable, perhaps unknowable, spaces and analytics of government that hardly correspond to the well-oiled machine of simply-theorised disciplinary biopower' (Staddon, 1998, p. 365).

What then does this mean for committed and progressive scholarship on water security? First, we contend that securing safe quantities and qualities of water (and the services it provides) for people and the environment is the only ethically acceptable perspective, and one that stands in contrast to realist environmental politics. Our scholarship must be about better understanding the challenges that real people in real communities are facing every day to secure safe water, and the often difficult choices they are forced to make to do so. Thus, we believe strongly in the need for detailed empirical fieldwork into the multiple dimensions and experiences of water security in multiple geographical contexts. This means also that we are interested in the underlying dynamics of water insecurity, particularly at the household scale (Jepson et al., 2017). Finally, it means that we are as interested, or perhaps more interested in, issues of water governance than we are in the natural science of climate change or population growth, for the root cause of environmental crisis is to be found not in how humans interact with nature, but in how they interact with each other. As Barry Commoner (1974, p. 32) put it in *The Closing Circle*: 'The debt to nature which is the measure of the environmental crisis cannot be paid person by person, in recycled bottles or ecologically sound habits, but in the ancient coin of social justice.'

Water security – the preferred discourse of global institutions

Despite the theoretical shortcomings of water security – or indeed perhaps because of its malleability and multiple adaptable meanings – the concept has been widely adopted by global and regional organizations spanning the multilateral, governmental and NGO spectrum. Early use of 'water security' accompanied the rise of broader human-security framings that sought to place access and control over water for human use (FAO, 1996) alongside ecosystem needs and ecological flows (UNEP, 2009). The Millennium Development Goals, focused on targets to fill the supply gap, were clearly predicated on human needs. As multilateral organizations of the United Nations and private-sector groupings increasingly employed water security discourse, countries, too, adopted the conceptual approach emphasizing access, equity, and ecological compatibility (though perhaps in ways that are different from the temporal, intergenerational sustainability of the Brundtland era).

Soon to jump on board the water security bandwagon were nonstate actors with global reach, including the World Water Council, the Global Water Partnership and the World Economic Forum. It is noteworthy that particularly the latter, representing business and commercial interests that extend far beyond those corporations that trade in water as a commodity (e.g., Coca Cola and Nestlé) or water services (e.g., Veolia), have seen water security as an attractive framing. Similarly, other UN bodies (UNESCO), regional groupings such as the Gulf Coordination Council and the Organization of American States, and universities (not least our own, the University of the West of England and the University of Arizona) have programatically enshrined water security.

Whither water security?

Water security, set as a target (a condition to be achieved) and a process (a capacity to be developed), serves multiple institutional interests. Yet, the conceptual appeal of water security must be tempered by limits to its application in operational terms. What does it mean to be water secure? Having achieved a state or condition of water security (if this is indeed definable), what constitutes a loss or reduction of water security? Will water security in the world of the Sustainable Development Goals and the Paris Climate Accord actually contribute to human and ecological outcomes? Reflecting on these and related *what-comes-next?* questions, the articles in this issue underscore our conviction that water security will continue to evolve, taking on new dimensions of social and environmental justice, access and outcomes, as well as offering a forward-looking perspective on the central role of water in broader environmental governance. Indeed, water security has already begun to morph into broader conceptions of human and planetary security.

Overview of articles in this special issue

With this conceptual and institutional backdrop, we as researchers and programme leaders who identify with water security themes approached the editor-in-chief of *Water International* with a proposal that has led to the present special issue. Here we briefly characterize the most salient findings and assertions of the articles that follow.

Wade addresses water security through a water-justice lens. Under a broader understanding of environmental justice, the article assesses (in)equity in resource distribution, diversity of stakeholders and communities, participation in decision making, and political power dynamics. Applying the analytical framework to the case of the Lake Naivasha cut-flower industry in Kenya, Wade demonstrates that governance structures can be co-opted by powerful commercial interests – which raises her central question of whether water justice is an elusive concept.

Nkhata considers water security from the perspective of benefit sharing, posing the question of whether the allocation of water and associated benefits is well suited to address local and basin-scale conflicts inherent in water security initiatives. Also taking a case-study approach (focusing on the Zambezi River basin), he shows that for benefit sharing to be successful, it must take account of and build on long-term relationships among multiple interest groups, often with divergent goals. It is only through such partnerships that water security can be a framing goal to share benefits.

Scott et al. seek to synthesize human- and resource-security understandings with a separate and not always compatible framework: the water–energy–food nexus. Tracing the historical emergence of food security and energy security, they place water security as the thread that interconnects the others at the level of resources, institutions, and multiple securities.

Albrecht et al. switch geographical scales to consider the difficult politics of transboundary water. Drawing on case studies from across the Americas, they argue that, 'Given the complexity introduced when political borders transect resource systems, in addition to the highly place-specific nature of water management, we expect to find a range of multidimensional water-security challenges and an equally broad array of institutional responses.' The case studies presented are richly drawn and intrinsically interesting, but the key central message of their analyses is that 'the chief compounding factors related to the transboundary context are national-sovereignty concerns and uneven power dynamics; political, economic, and physiographic asymmetries; and insufficient institutional capacity.' In other words, there is a burgeoning challenge to find new governance mechanisms, not based on the zero-sum logics of Bismarckian nation-states, to better manage water across boundaries.

Staddon et al. present research findings that illuminate the household decision-making dynamics of adopters and non-adopters of a water collection and storage technology. Through survey research with households in three districts in central Uganda, the authors discovered that the most important determinants of the decision to adopt ferro-cement rainwater harvesting technologies is age of head of household and membership in or affiliation with a promoting organization. With this greater understanding of adoption trends and drivers, the authors contend that governmental and non-governmental organizations may be able to make more informed decisions on where to focus their attempts to better support communities or households that may previously have been marginalized.

Notes

1. Social Darwinism refers to the notion that societies are shaped by their physical environments, to which they adapt, just as species adapt.
2. The Washington Consensus was formed in the late 1980s to promulgate the idea that only private property transacted in a free market with minimal government interference could result in rational allocation of resources and the creation of robust economies.

Disclosure statement

No potential conflict of interest was reported by the authors.

Funding

This work was undertaken with partial support of the International Water Security Network, a project funded by Lloyd's Register Foundation, a charitable foundation helping protect life and property by supporting engineering-related education, public engagement and the application of research. Additional support was provided by the Inter-American Institute for Global Change Research (project CRN3056), which is supported by the US National Science Foundation (grant no. GEO-1128040), NSF grant no. DEB-1010495, and the Morris K. Udall and Stewart L. Udall Foundation in Tucson, Arizona.

ORCID

Chad Staddon http://orcid.org/0000-0002-2063-8525

References

Afolayan, A. A., & Adelekan, I. O. (1999). The role of climatic variations on migration and human health in Africa. *The Environmentalist, 18*(4), 213–218.

Commoner, B. (1974). *The closing circle*. New York: Knopf.

Cook, C., & Bakker, K. (2012). Water security: Debating an emerging paradigm. *Global Environmental Change, 22*, 94–102.

Diamond, J. (2005). *Collapse: How societies choose to fail or succeed*. New York: Viking Books.

Ezra, M. (2001). Demographic responses to environmental stress in the drought- and famine-prone areas of northern Ethiopia. *Population, Space and Place, 7*(4), 259–279.

Febvre, L. (1932). *A geographical introduction to history*. London: Kegan Paul, Trench, Treabner (originally published in 1922 as La terre et l'évolution humaine).

Food and Agriculture Organization (FAO). (1996, November 13–17). *World food summit plan of action*. Rome, Italy: World Food Summit.

Forsyth, T. (2003). *Critical political ecology: The politics of environmental science*. London: Routledge.

Gerlak, A. K., House-Peters, L., Varady, R., Albrecht, T., Zuniga Teran, A., de Grenade, R., … Scott, C. A. (2018). Water security: A review of place-based research. *Environmental Science & Policy, 82*, 79–89.

Grey, D., & Sadoff, C. W. (2007). Sink or swim? Water security for growth and development. *Water Policy, 9*(6), 545–571, 27.

Homer-Dixon, T. (1994). Environmental and demographic threats to Canadian security. *Canadian Foreign Policy, 2*(2), 7–40.

Jarosz, L. (2004). Political ecology as ethical practice. *Political Geography, 23*(2004), 917–927.

Jepson, W., Budds, J., Eichelberger, L., Harris, L., Norman, E., O'Reilly, K., … Young, S. (2017). Advancing human capabilities for water security: A relational approach. *Water Security, 1*, 46–52.

Klare, M. T. (2001). *Resource wars: The New Landscape of global conflict*. New York: Metropolitan Books.

Liverman, D. (2009). Conventions of climate change: Constructions of danger and the dispossession of the atmosphere. *Journal of Historial Geography, 35*(2), 279–296.

Luke, T. W. (1999). Environmentality as green governmentality. In E. Darier (Ed.), *Discourses of the environment* (pp. 121–151). Malden, PA: Blackwell.

Mouffe, C. (2005). *On the political*. London: Routledge.

Myers, N. (1994, March/April) Ultimate security: The environmental basis of political stability, *Foreign Affairs*

Ohlsson, L. (1995). *Water and security in Southern Africa*. Stockholm: Department for Natural Resources and the Environment, SIDA.

Peace, A. (2002). Governing the environment: The programs and politics of environmental discourse. In C. O'Farrell (Ed.), *Foucault: The legacy, proceedings of the Foucault: The legacy conferenceheld in Australia, (1994)* (pp. 530–545). Kelvin Grove, QLD: University of Queensland.

Schulz, M. (1995). Turkey, Syria and Iraq: A hydropolitical security complex. In L. Ohlsson (Ed.), *Hydropolitics: Conflicts over water as a development constraint* (pp. 20–35). London: Zed Books.

Shuval, H. (1992). Approaches to solving water resources conflicts in arid areas–Israel and her neighbors as a case study. In *Legal issues in water resources allocations, waste water reuse, and water management* (pp. 24–46). Geneva, Switzerland: World Health Organization.

Staddon, C. (1998). "Democratisation and the politics of water in Bulgaria: Local protest and the 1994-5 Sofiya Water Crisis." In J. Pickles & A. Smith (Eds.), *Theorising transition: The political economy of postcommunist transformations* (pp. 373–388). London: Routledge.

Staddon, C., & James, N. (2014). Water security: A genealogy of emerging discourses. In G. Schneier-Madanes (Ed.), *Globalised water: A question of governance* (pp. 261–276). Dordrecht: Springer.

United Nations Environment Program (UNEP). (2009). *Water security and ecosystem services: The critical connection.* Nairobi: UNEP.

Varady, R. G., Albrecht, T. R., Gerlak, A. K., Zuniga, A. A., & Staddon, C. (in press). The water security discourse and its main actors. In J. J. Bogardi, K. D. Wasantha Nandalal, R. R. P. van Nooyen, & A. Bhaduri (Eds.), *Handbook of water resources management.* Dordrecht: Springer

Is water security just? Concepts, tools and missing links

Sarah Wade

ABSTRACT

Both social and environmental justice overlap with water (in)security, but neither fully captures the nuances. This review extends a water justice framework by critiquing and further developing an existing environmental justice framework. Testing a reformulated understanding of water security through a case study reveals added insights derived from inclusion of scale and power dynamics.

Introduction

The notion of justice, be it social or environmental, is a popular theme in research around allocation and use of natural resources as we seek to answer questions about why particular decisions or actions are taken, and how these affect diverse stakeholder groups. Globally, we aim to 'ensure availability and sustainable management of water and sanitation for all' (Sustainable Development Goal 6) by 2030, and while progress has been made, we still have a long way to go. 'Justice' provides a lens through which we can explore the reasons why some communities still lack access to clean, safe water, or why certain powerful actors make water-related decisions.

Some have criticized the use of justice framings for water security challenges. McLean (2007) and Strang (2005) argue that existing social and environmental justice frameworks are unable to capture the diversity and complexity of water security. Strang highlights that water is universally required by all, but societies' relationships with water vary across contexts, shaped in part by our diverse methods of valuing water. McLean argues that the cultural value of water is often overlooked, as is the complex role of the private sector as a water provider when public utilities do not exist. Patrick, Syme, & Horwitz (2014) highlight four characteristics of water with implications for justice: the spatially and temporally uneven distribution of water; the fact that water is essential for all life, with minimum requirements for both people and the environment; the benefits that water creates for well-being through the goods and services it produces; and the political dimensions of power asymmetries affecting water governance.

This article critiques and modifies an existing framework for environmental justice in response to challenges identified in accurately representing water security through the lens of environmental justice. It builds on an existing environmental justice framework to create a matrix (Table 1) for analyzing water (in)justice from a water security perspective. Approaching (in)justice through the lens of water security provides insight into the

Table 1. A matrix for use when applying the framework.

Levels (jurisdictional scale)	Equity in distribution (of water resource)	Recognition of diversity of communities	Participation in decision-making process	Power dynamics (hard and soft; 1st, 2nd and 3rd dimensions)
Local				
National				
International				

processes creating water (in)justice, including the coexistence of multiple justices and injustices. The framework is applied to a case study of Kenya's cut-flower exports, in view of the crucial role of this industry in water use and (in)justice in Kenya.

Justifying justice: concepts, tools, and the missing links

A literature review is used to examine an existing environmental justice framework from a water security perspective to identify modifications required to develop it into a framework tailored for application to cases of water (in)justice. The review begins with an overview of water security, the conceptual lens through which justice is viewed throughout the article. It goes on to explore the literature on environmental and water justice to identify the fundamental requirements of a water (in)justice framework. The review concludes with the relevant literature on scale and power, as these are shown in the preceding sections to be vital considerations in relation to decision-making around water, but currently missing from environmental justice frameworks.

Water security

The Global Water Partnership (2000, p. 12) provided one of the first definitions of water security: 'Water security at any level from the household to the global means that every person has access to enough safe water at affordable cost to lead a clean, healthy and productive life, while ensuring that the natural environment is protected and enhanced.' While many definitions mirror this, others have a different focus, for example competing livelihoods (Grey & Sadoff, 2007), privatization (Bakker, 2003), water for agriculture, energy production or the environment (Rosegrant, Ringler, & Zhu, 2009; Turral, Burke, & Faurès, 2011; Vörösmarty et al., 2010), or water for basic needs (Sullivan, 2002). What is consistent across them all is that water security is underpinned by many of the same components as social justice: both are based on the notion that everyone deserves equal rights and opportunities, whether economic, social or political. Consequently, a water security perspective inherently includes a social justice viewpoint. This article uses Grey and Sadoff's (2007, p. 548) definition of water security – 'the availability of an acceptable quantity and quality of water for health, livelihoods, ecosystems and production, coupled with an acceptable level of water-related risks to people, environments and economies' – as it provides a holistic viewpoint that is relevant in multiple contexts.

Environmental justice

While originally separate, environmental justice is now broadly accepted as a combination of distributional and procedural justice. Distributional justice considers how environmental costs and benefits fall disproportionately on certain communities (Domènech, March, & Saurí, 2013), while procedural justice focuses upon the decision-making process around these actions, with consideration of who is not included in the process and the effects this exclusion may have (Domènech et al., 2013).

More recently, environmental justice has been expanded to include three core elements: distribution, participation and recognition (Holifield, Porter, & Walker, 2009; Schlosberg, 2004). Schlosberg (2004, p. 521) says that 'justice demands a focus on recognition, distribution and participation', calling them 'three interlinking, overlapping circles of concern'. Based on this, he creates a framework for environmental justice which incorporates equity in distribution of risk, recognition of diversity of participants and experiences in affected communities, and participation in political processes which create and manage policies. Schlosberg argues that this framework avoids a common over-focus on distributional justice which limits consideration of recognition and participation. Young (1990) also highlights this and argues for a social justice perspective to be included with environmental justice to encourage consideration of recognition and participation, which in turn creates a condition whereby the factors underpinning poor distribution can be considered. Equity is included by Schlosberg within the framework; this concept is linked with justice, and sometimes there is confusion between the two. Distributive justice (or equity in distribution, as Schlosberg calls it) is what is often meant by 'equity'. Ikeme (2003) argues that environmental justice is an overarching concept which includes both procedural and distributional justice (equity), but Schlosberg goes further by incorporating recognition as well. The links between the notions of equity and justice, and their interaction in relation to water, warrant further research.

As a concept, environmental justice has received some criticism. Simply put, it can lead to scenarios being described from a 'heroes and villains' viewpoint which fails to consider case-specific factors (Williams & Mawdsley, 2006). This can hamper investigation of the root cause of injustice, as the researcher may overlook spatial and social contexts which determine relationships between stakeholders (Holifield et al., 2009). Power is not considered in traditional framings of environmental justice, and yet in water security, power has been demonstrated to be a strong shaper of outcomes in relation to access to and control over water (Domene, 2006; Zeitoun & Allan, 2008). Another potential weakness of environmental justice is described by Domènech et al. (2013, p. 23), who argue that it 'has been traditionally focussed on the local scale while glossing over the multi-scalar processes and relations contributing to the socio-ecological inequalities and injustices at the local scale'. Both power and scale play clear roles in water security and water justice.

Sikor (2013) and Sen (2009) draw on a multidimensional approach to justice to demonstrate that we can no longer consider there to be 'universally shared notions of justice' (Sikor, 2013, p. 7) and that in some situations multiple 'just' or 'unjust' outcomes may be created and occur simultaneously (Sen, 2009).

Schlosberg's framework has several merits for investigating environmental justice, including water justice. It has already been used to consider cases of water (in)justice, including by McLean (2007) and Mascarenhas (2007). McLean argues that application of Schlosberg's framework to water b greater recognition of community water rights than other framings of water justice while avoiding the 'heroes and villains' scenario. Based on these positive findings, Schlosberg's framework will be used as the basis for a water justice framework. Considering the framework within the context of water security, which shares its roots with social justice, roots it more in shared notions of both environmental and social justice. The three core elements of Schlosberg's framework will be retained, but additions will be made to address the identified shortfalls for application to water-related challenges. The following sections consider the relevant literature to build out Schlosberg's framework into a framework tailored for application to water.

Water justice

To date, the most comprehensive reviews of 'water justice' have been provided by Patrick (2012, 2014), Patrick, Syme, & Horwitz (2014), Patrick et al. (2014), McLaughlin (2013), and Zwarteveen and Boelens (2014). McLaughlin's literature review notes that despite water being tied to much of the literature on social and environmental justice through the challenges it creates, there are several gaps in conceptualizations of 'water justice'. This is consistent woth calls from others who argue that environmental justice framings alone are not appropriate in considering cases of water (in)justice (McLean, 2007; Strang, 2005).

Zeitoun and McLaughin (2013) argue that in water justice, there are trade-offs between three factors which can complicate the narrative: between users, uses, and biophysical processes. These competing factors are also demonstrated in the range of water security definitions described previously, further demonstrating the intrinsic links between justice and water security.

Patrick (2012, 2014) developed a framework for water justice called Cycles and Spirals of Justice, which combines social and environmental justice frameworks with a landscape-ecology understanding of scales and levels. The framework highlights the continuum of justice and injustice across multiple scales and levels in transboundary water resource management; a just outcome at one level could create or exacerbate injustices at another. Further work by Patrick et al. (2014) has shown that procedural justice can influence stakeholders' likelihood of accepting a decision or outcome, as it can enhance the perception that the process of reaching the decision or outcome has been fair. They note that soft power can further entrench existing unfair outcomes in water resource management. However, the concept of power is not explicitly covered by Cycles and Spirals of Justice.

Zwarteveen and Boelens (2014) suggest several concepts and tools for use in water justice research and action. Any tool must recognize the power and politics behind water use, management and governance, reflecting calls from McLean (2007), Schlosberg (2004), and Zeitoun (2008). Zwarteveen and Boelens (2014) highlight the vital consideration that must be given to invisible power in water dynamics. Fundamentally, they argue that we must not focus purely on what 'should' be, but on how injustice is experienced. They describe water justice as 'intrinsically scale-sensitive' (p. 151), reiterating the need to include scale in justice research.

The missing links: scale and power

Scales (and levels)

In much of the literature reviewed, Cash et al.'s (2006) definition of scales and levels is used, in which scales constitute spatial, temporal, jurisdictional or institutional contexts, and levels account for the points along each of the scales, for example the 'local' or 'national' level on the jurisdictional scale. Patrick (2012) has used this conceptualization with water justice, although other writers use the terms slightly differently, with some using 'scale' to refer to multiple levels (Marston, Jones, & Woodward, 2005).

Zeitoun (2013, p. 143) argues that 'A disparate group of water justice articles have remarked on the dissonance/mismatch between the geographical scale at which the environmental injustices are experienced (i.e., "local"), and the (global) scale at which they are produced.'

Blaikie and Brookfield (1987) call for a scalar approach in work with natural resources, highlighting the impact of hierarchical socio-economic organizations on natural resource management outcomes. While Zeitoun (2013) has noted the risk in an over-focus on the 'global' perspective, Purcell and Brown (2005) highlight a 'local trap': there can be an assumption that localized decision-making is inherently more just, something which is not always the case. They argue that scales (and levels) are dependent on political agendas. As a result, it is key to this research that both scales and levels only provide a part of the picture of water justice, to avoid the risk of over-focusing on any one element.

Consideration of multiple scales creates a duty to consider multiple institutions, approaches, perspectives and priorities, including consideration of power, which influences the connections between scales and levels. Scale is clearly a vital addition to Schlosberg's framework for application to water, albeit with caution, as mentioned. It is Cash et al.'s definition of scales and levels that will be used in the framework, as its strength has been demonstrated through Patrick's application to water justice. However, with each application of the framework, the specific scale and related levels should be redefined to best reflect the case under consideration. It is important to ensure that the choice of scale(s) does not create bias towards particular stakeholders (Lebel, 2006).

Power

Most research on the relationship between water and power has focused on the context of international rivers, drawing on power framings from Nye (2003) and Lukes (2005 [1974]). Zeitoun (2013) argues that failure to fully consider the role of power dynamics in relation to water can result in a breakdown of the key Rawlsian condition of justice being delivered under a 'veil of ignorance'.

Lukes (2005 [1974]) built on work by Dahl (1957) and Bachrat and Baratz (1962) to develop the three dimensions of power, which were used by Zeitoun and Warner (2006) in their work on hydro-hegemony. The first dimension (included in Dahl's original work) is sometimes called *material power*; it focuses on power in decision making: 'A has power over B to the extent that he can get B to do something that B would not otherwise do' (Dahl, 1957, p. 202). The second dimension draws on the mobilization of bias and the role of *non-decision-making*, which according to Bachrat and Baratz (1962) is the process which prevents issues from reaching the policy arena. Lukes describes this as the ability to set the agenda. Lukes developed the third dimension of power in response to criticisms of earlier work, aiming to consider the role of collective action and the absence of visible conflict. This dimension is known as *thought control*, where stakeholders are prevented from realizing that a concern exists, thereby excluding it from the political arena.

Nye (2003) provides an alternative conceptualization of power focused on 'hard' and 'soft' power. 'Soft power is the ability to get what you want by attracting and persuading others to adopt your goals. It differs from hard power, the ability to use the carrots and sticks of economic and military might to make others follow your will' (p. 1). There are clear links between Lukes's first dimension of power and Nye's hard power, and again with the second and third dimensions alongside soft power.

While both conceptualizations of power have received criticism (Xiaoming, 2005), they have also been used extensively in relation to water. Zeitoun and Mirumachi (2008), followed by Zeitoun, Mirumachi, and Warner (2011), investigated the role of soft power in influencing conflict and cooperation in transboundary river contexts. Zeitoun and Allan (2008) highlighted the potential soft power has in relation to water, particularly due to the invisible elements.

Power has already been identified as a missing link in Schlosberg's framework when applied to water, and it is clear from the literature that both Lukes's and Nye's conceptualizations have merits which warrant inclusion in the framework. In many cases, all three dimensions (or both hard and soft power) may not occur simultaneously, or there may be multiple dimensions occurring at every level. The factors driving power dynamics must also be considered when the power conceptualization is chosen and applied, as some cases may lend themselves more holistically to one or the other. Both conceptualizations are given as options in the framework, but neither is expected to be used exhaustively, and both may be used simultaneously.

A framework of water justice

The framework developed based on the critique of existing approaches takes the form of a matrix (Table 1). The column headings are fixed, but the scales and levels depend on the case study, so before applying the framework, the researcher must set the scale(s) and levels based on Cash et al.'s definition, as noted previously.

Application of the framework through completion of the matrix must be based on a comprehensive understanding of the context within which the water issue exists. The following four metrics can be used to help identify relevant information.

Equity in distribution

Equity is a complex notion which can be viewed through many lenses, but in this case, due to the link to 'distribution' of water resources, it can be considered through the economic lens of a cost/benefit analysis in relation to water security. For example, at each of the defined levels, who bears the costs and who enjoys the benefits of adequate access to clean water? Do the costs (financial, time, cultural) prevent a stakeholder group or community from being water-secure? Is one group's enjoyment of water-related benefits occurring to the detriment of another's? Equity is a relative concept which can vary by context, so as different levels are considered, these questions should be reconsidered.

Recognition of diversity of communities

At each level, a thorough understanding of all stakeholder groups must be used to consider the extent to which different communities are recognized and empowered to achieve water security. Once stakeholders are identified, contextually specific metrics might be available to provide further insight into the challenges they face in relation to water. For example, the average amount of time spent collecting water per household, or provisions in local or national law for marginalized groups to be included in the decision-making process, might help identify the level of recognition provided to different communities.

Participation in the decision-making process

This section of the matrix relies heavily on qualitative and potentially anecdotal evidence. While legal provision may exist for participation in decision-making across diverse communities and at various levels, this may not be realized in practice. National reporting provides insight into intentions (for example through designation of funds to support local water user associations), but testing how this plays out across the levels will identify the extent to which participation is realized.

Power dynamics

Depending on the case, different conceptualizations of power may provide greater insight. Where limited stakeholder groups or levels are involved, power dynamics might be more obvious and hard and soft power provide a sufficient level of detail. In more complex cases, the three dimensions of power may be required to fully identify all the influential factors. The information can be used to identify the 'most powerful' actor or stakeholder group; this could aid identification of a driver or process creating injustice.

Applying the framework: Lake Naivasha, Kenya

As noted in the introduction, the framework is tested through application to the case of Lake Naivasha in Kenya. According to the Sustainable Development Goals Report (United Nations, 2018), in 2015, 29% of the world's population lacked access to safely managed drinking water supplies, and in 22 countries water stress was over 70%, indicating high probability of future water scarcity. In Kenya, 37% of the population is reliant on an unimproved source (WHO/UNICEF, 2015), but the country is also a major exporter of a variety of produce known to be water-intensive in it production or to have a detrimental impact on water quality. For example, Kenya provides 38% of the cut-flower imports to the EU every year (Kenya Flower Council, 2016), and the cut-flower industry is largely located in a region suffering from challenges in both water availability and water quality.

Several scales could be chosen at which to apply the framework. It is important that the scale represents the nuances of the case, while providing space for full consideration of the wider landscape. In Kenya, the role of international stakeholders, coupled with local reliance on Lake Naivasha, lends itself to the jurisdictional scale, with levels set at local, national and international. It is important to note that from a water perspective, local means the sub-catchments and national encompasses both the catchment level and national-government involvement. There is also a risk in the need to define scales and levels; if these are incorrectly identified, or too closely tied to social constructs and assumptions, they may not accurately represent the case.

The completed matrix for Lake Naivasha is shown in Table 2.

Equity in distribution

The first element to consider is equity in distribution. Locally, there appears to be fundamental injustice, with many stakeholders lacking access to water, particularly smallholders, fishermen and domestic users who do not have a piped supply. Taken further, additional injustice can be found in payment for water, as poor regulation means that many unregistered users do not pay for their water use, while registered users cover the costs (Mekonnen & Hoekstra, 2010). This injustice is felt more keenly by the commercial users of the lake.

At the national level, the intensity of injustice is less as the lake provides several economic, social and infrastructure-based opportunities and gains (Mekonnen & Hoekstra, 2010). Internationally, a divergent form of injustice is identified in relation to allocation; despite only a small proportion of the produce from Naivasha being sold internationally, the commercial farms account for 87% of Lake Naivasha's blue-water footprint.[1]

Recognition of diversity of communities

Scale underpins this element of the framework. Locally, recognition seems to favour agricultural water users and those who are officially registered to use the lake, but as we move to the national level, the establishment of water resource users' associations attempts to rectify the balance, as the associations are intended to represent all users equally (Bolo, 2008). The detail here highlights a process whereby water injustice is being created due to weak regulations and a lack of capacity to impose legal requirements, meaning that some users are not represented, despite the mechanisms intended to ensure this (WWF International, 2012).

Participation

In the case of Naivasha, participation appears to be dominated by water users associated with flower production for international markets. This is shown to be the case at both the local and international levels, where flower growers are well represented in associations whose core purpose is to ensure their participation in decision-making (Bolo, 2008). International and local NGOs work with local smallholders and growers' groups to ensure their participation in decisions taken at the national level (WWF International, 2012).

Table 2. Completed framework of water justice matrix for Lake Naivasha, Kenya.

Levels (jurisdictional scale)	Equity in distribution (of water resource)	Recognition of diversity of communities	Participation in decision-making process	Power dynamics (hard and soft; 1st, 2nd and 3rd dimensions)
Local (sub-catchment)	Farmers, fishermen, herdsmen and domestic users sometimes struggle to access the lake. Migrant-worker settlements are reliant on water resources from the lake. Commercial farmers argue that smallholders in the upper catchment are unregistered water users and therefore don't pay. (There are also many unregistered users around the lake – Mekonnen & Hoekstra, 2010).	Farmers, whether small or large-scale, appear to dominate water use, followed by people in the towns who have piped water coverage which relies on the lake. More 'traditional' users seem to have little recognized claim to water. The Water Services Regulatory Board is meant to ensure a minimum quality of drinking water, but many local people take water directly from the lake.	Official groups (Kenya Flower Council, Lake Naivasha Growers Group, Lake Naivasha Riparian Association) are meant to represent farmers in the decision-making process. Other stakeholders (in the upper catchment, and non-agricultural users of the lake) are not as well represented by groups, and therefore often left out of the decision-making process.	Multiple narratives at the local level. Commercial farmers (and others) argue that since they pay for their water use, while many unregistered users in the upper catchment do not, they should receive some benefits from the system. Subsistence farmers and some smallholders argue that commercial farmers are responsible for the decreasing water level and quality. These (and many other narratives) could represent various users of the second dimension of power.
National	Water used to generate foreign income and for drinking water supply in areas with coverage in Nakuru and Naivasha Towns. Also vital for electricity generation.	Under the Water Act of 2002, the Ministry of Water and Irrigation is responsible for water use. It established the Water Resource Management Authority, which in turn established water resource users' associations (WRUAs), which are meant to include all water users: domestic, commercial farmers, fishermen, etc. (Bolo, 2008).	Under the Water Act, there is a tiered system to manage water resources. Some international NGOs are working with the Ministry of Water and Irrigation to increase the effectiveness of WRUAs, which could increase participation of multiple stakeholder groups (Bolo, 2008).	The Lake Naivasha WRUA is well developed and organized around multiple existing groups and associations. This means that it can dominate discussions around water use at the lake, as other WRUAs, also based in the Naivasha basin, are less developed, less well represented, and therefore less powerful. Again, this represents use of the second dimension of power.
International	Agribusiness dominates water use. Although around 90% of agricultural produce from around the lake is sold locally, the commercial farms that export most of their produce account for 87% of the blue-water footprint (WWF International, 2012).	There is heightened consumer awareness of the impacts of farming around Naivasha, which has the potential to impact the industry. Many supermarkets and international buyers of flowers produced in Kenya are actively trying to improve their image with consumers.	Some larger companies, particularly the international ones, might be able to use soft power to ensure their involvement in decision-making processes. Due to the international nature of horticultural trade, larger international businesses are better placed to take part in international discussions of standards and policies.	International agribusinesses use hard power to maintain their water use; they threaten to relocate to other countries if the price of water is raised (Bolo, 2008). Some soft power is at work, as businesses claim to provide infrastructure for local people, thereby manipulating local and national government, and potentially protecting their corporate image.

Source: Contextual information from Becht, Odada, & Higgins (2005), Bolo (2008), Enniskillen (2002), and Mekonnen and Hoekstra (2010).

Power

The second dimension of power that emerges from this case comes to light locally and internationally, where competing narratives are promoted. This highlights a strength of the multilevel approach of the framework, as consideration of power dynamics at just one level could lead to singular, level-based narratives adopted inaccurately.

The third dimension of power is evident in the construction and maintenance of the dominant narrative by particular stakeholder groups in Naivasha. Western media has highlighted flower farming as particularly detrimental to local water security and conservation, and this narrative has been echoed by development and conservation organizations. This may demonstrate power in the third dimension influencing narrative creation and uptake, but it would be hard to prove.

At the international level, large-scale agribusinesses have threatened to leave Kenya if water prices rise, so it can be argued that hard power is at play (Mekonnen & Hoekstra, 2010). Large businesses leaving would hurt Kenya's economy and local livelihoods, highlighting the power held by the international agribusinesses.

A key learning from applying this element of the framework is that at each step, the visible and invisible power dynamics must be considered and noted. Limitations of the framework will arise if generalizations are made, as details may be lost that should be drawn out through consideration of power dynamics and may be useful in creating an accurate narrative of the (in)justice.

What does application of the framework tell us about water justice in Kenya?

Applying the framework has provided a multilevel understanding of the justice and injustice which coexist around Kenya's cut-flower industry and the broader issues of water security around Lake Naivasha. The inclusion of scale in the framework has enabled exploration of competing narratives on water use through different stakeholders' perspectives, something which existing frameworks of environmental justice do not sufficiently consider in their application to water.

The costs and benefits of the cut-flower industry and its associated water use are felt differently across society. At the national level, it could be argued that the economic benefits of the industry outweigh the costs, as the benefits are felt by the national economy and the costs are (predominantly) felt at a local level; the impact of the water use is most acutely felt by local communities reliant on the lake for subsistence.

The framework demonstrates that while there are governance structures in place in Kenya to provide opportunities for participation in decision-making, these can be dominated by some of the larger stakeholder groups, or by those nearest to the lake, leaving marginalized or more rural communities with little power to influence decision making.

Consideration of power dynamics provides further insight into the process creating and perpetuating injustice. For example, the role of the second dimension of power can be seen in the heavy influence enjoyed by the more established associations representing the commercial farmers.

Water justice: elusive concept or useful tool?

Scale and power have intrinsic links to the concept of water justice. The framework combines the strength of an environmental justice approach with the detail of a multidisciplinary water security perspective, linking the roles of scale and power to provide greater detail on narratives surrounding water (in)justice.

Scale provides the opportunity to take a holistic view of a situation, incorporating a variety of perspectives and viewpoints to build a multilevel understanding. When acting on injustice, it is vital to have a full understanding of relationships between justice and injustice at various levels; without this understanding, action could be taken to mitigate an injustice at one level which inadvertently exacerbates injustice at another level. A multiscalar approach highlights opportunities for cooperation to resolve injustices across a variety of economic, political and social contexts.

Patrick's (2012) work on the Cycles and Spirals of Justice highlights the interaction between justice and injustice across various levels within multiple scales to form a useful tool for research on cases of water (in)justice. An approach such as this combined with the additional detail of the framework developed here provides a strong research base on water injustice combined with clear viewpoints on potential resolutions to the challenge.

The addition of power to the framework provides a detailed lens through which to investigate causes of injustice, enabling a comprehensive understanding of visible and invisible factors that can perpetuate water injustice. Separate consideration of power dynamics draws attention to the process behind an injustice, mitigating a core concern, identified in the literature review, that spatial and social contexts can be overlooked by one-dimensional frameworks.

Application of the framework demonstrated the need for water justice to have a strong understanding of the context from multiple perspectives to avoid overlooking any factors which may create or contribute to (in)justice. The perspective of the stakeholders involved may influence representation of the issue, something which should be mitigated through consideration of power dynamics and narrative creation, combined with an understanding of stakeholders' and researchers' proximity to the issue.

This water justice framework is a useful and powerful tool for research and action on water (in)justice. It challenges static conceptualizations of justice, identifying the potential coexistence of justice and injustice, and encouraging research and action to view justice as a process, not simply an outcome. A key strength of the framework is its ability to draw in the perspective of stakeholders across a variety of levels and consider the power dynamics influencing their engagement with water. This demonstrates its potential suitability as a tool for examining stakeholder conflict and undertaking stakeholder mediation to remedy issues of water injustice.

Note

1. 'A water footprint captures both the blue and green components of water consumption. A blue water footprint refers to the volume of surface and groundwater that is used for irrigation in the supply chain of a product (net abstraction less return flow), while the green footprint refers to the consumption of rainwater that is evapotranspirated from soil moisture' (WWF International, 2012, p. 15).

Disclosure statement

No potential conflict of interest was reported by the author.

References

Bachrat, P., & Baratz, M. S. (1962). Two faces of power. *American Political Science Review, 56*, 941–952.

Bakker, K. (2003). Archipelagos and networks: Urbanization and water privatization in the South. *The Geographical Journal, 169*(4), 328–341.

Becht, R., Odada, O., & Higgins, S. (2005) *Lake Naivasha: Experience and lessons learned brief.* Retrieved from https://research.utwente.nl/en/publications/lake-naivasha-experience-and-lessons-learned-brief

Blaikie, P., & Brookfield, H. (1987). *Land Degradation and Society*. New York: Methuen.

Bolo, M. O. (2008). The Lake Naivasha Cut Flower Cluster in Kenya. In D. Z. Zeng (Ed.), *Knowledge, Technology and Cluster Based Growth in Africa* (pp. 37–52). Washington, DC: The World Bank.

Cash, D. W., Adger, W. N., Berkes, F., Garden, P., Lebel, L., Olsson, P., … Young, O. (2006). Scale and cross-scale dynamics: Governance and information in a multilevel world. *Ecology and Society, 11*(2), 8. Retrieved from http://www.ecologyandsociety.org/vol11/iss2/art8/

Dahl, R. A. (1957). The Concept of Power. *Behavioural Science, 2*(3), 201–215.

Domene, E. (2006). La ecología política urbana: Una disciplina emergente para el análisis del cambio socioambiental en entornos ciudadanos. *Documents d'Anàlisi Geogràfica, 48*, 167–178. Retrieved from http://www.raco.cat/index.php/DocumentsAnalisi/article/viewFile/72658/82968

Domènech, C., March, H., & Saurí, D. (2013). Contesting large-scale water supply projects at both ends of the pipe in Kathmandu and Melamchi Valleys, Nepal. *Geoforum, 47*, 22–31.

Enniskillen, A. (2002). Introduction: The Lake Naivasha Management Plan–Consensus building to conserve an international gem. *Hydrobiologia, 488*(168), 9–12.

Global Water Partnership. (2000). *Towards water security: A framework for action*. Stockholm, Sweden: Author.

Grey, D., & Sadoff, C. (2007). Sink or swim? Water security for growth and development. *Water Policy, 9*(6), 545–571.

Holifield, R., Porter, M., & Walker, G. (2009). Introduction: Spaces of environmental justice: Frameworks for critical engagement. *Antipode, 41*, 591–612.

Ikeme, J. (2003). Equity, environmental justice and sustainability: Incomplete approaches in climate change politics. *Global Environmental Change, 13*(3), 195–206.

Kenya Flower Council. (2016). Retrieved from http://kenyaflowercouncil.org/

Lebel, L. (2006). The politics of scale in environmental assessments. In W. V. Reid, F. Berkes, T. J. Wilbanks, & D. Capistrano (Eds.), *Bridging scales and knowledge systems: Concepts and applications in ecosystem assessments* (pp. 37–58). Washington DC: Island Press.

Lukes, S. (2005 [1974]). *Power: A radical view* (2nd ed.). Hampshire, UK: Palgrave MacMillan.

Marston, S. A., Jones, J. P., & Woodward, K. (2005). Human geography without scale. *Transactions of the Institute of British Geographers NS, 30*, 416–432.

Mascarenhas, M. (2007). Where the waters divide: First Nations, tainted water and environmental justice in Canada. *Local Environment: the International Journal of Justice and Sustainability, 12*(6), 565–577.

McLaughlin, K. (2013, January) Water and Justice: A literature review. *Presented at the Sixth Hydrohegemony Conference*, London.

McLean, J. (2007). Water injustices and potential remedies in indigenous rural contexts: A water justice analysis. *Environmentalist, 27*, 25–38.

Mekonnen, M. M., & Hoekstra, A. Y. (2010). *Mitigating the water footprint of export cut flowers from the Lake Naivasha Basin, Kenya. (Research Report Series No. 45)*. Delft: UNESCO-IHE Institute for Water Education.

Nye, J. (2003, January 10). Propaganda isn't the way: Soft power. *International Herald Tribune*.

Patrick, M. J. (2012) Justice and Scale in Water Allocation (Unpublished PhD Thesis). Faculty of Business and Law, School of Law and Justice. Edith Cowan University, Perth. Retrieved from http://ro.ecu.edu.au/theses/474/

Patrick, M. J. (2014). The cycles and spirals of justice in water-allocation decision making. *Water International*, *39*(1), 63–80.

Patrick, M. J., Komakech, H., Mirumachi, N., Moosa, H., Prakash, A., Salame, L., … Wolf, A. T. (2014). Building bridges between the sciences and the arts of water co-operation through collective action-reflections. *Aquatic Procedia*, *2*, 48–54.

Patrick, M. J., Syme, G. J., & Horwitz, P. (2014). How reframing a water management issue across scales and levels impacts on perceptions of justice and injustice. *Journal of Hydrology*, *519*, 2475–2482.

Purcell, M., & Brown, J. C. (2005). Against the local trap: Scale and the study of environment and development. *Progress in Development Studies*, *5*(4), 279–297.

Rosegrant, M. W., Ringler, C., & Zhu, T. (2009). Water for agriculture: Maintaining food security under growing scarcity. *Annual Review of Environment and Resources*, *34*, 205–222.

Schlosberg, D. (2004). Reconceiving environmental justice: Global movements and political theories. *Environmental Politics*, *13*, 517–540.

Sen, A. (2009). *The Idea of Justice*. London: The Penguin Group.

Sikor, T. (2013). *The justices and injustices of ecosystem services*. London and New York: Routledge.

Strang, V. (2005). Water works: Agency and creativity in the Mitchell river catchment. *Australian Journal of Anthropology*, *16*(3), 366–381.

Sullivan, C. (2002). Calculating a water poverty index. *World Development*, *30*(7), 1195–1210.

Turral, H., Burke, J., & Faurès, J.-M. (2011) *Climate change, water and food security* (FAO Water Report 36). Rome: Food and Agriculture Organisation of the United Nations.

United Nations. (2018) The Sustainable Development Goals Report 2018. Retrieved from https://unstats.un.org/sdgs/files/report/2018/thesustainabledevelopmentgoalsreport2018.pdf

Vörösmarty, C. J., McIntyre, P. B., Gessner, M. O., Dudgeon, D., Prusevich, A., Green, P., … Davies, P. M. (2010). Global threats to human water security and river biodiversity. *Nature*, *467*(7315), 555–561.

WHO/UNICEF Joint Monitoring Programme. (2015). *Progress on sanitation and drinking water: 2015 update and MDG assessment*. Geneva: WHO Press.

Williams, G., & Mawdsley, E. (2006). Postcolonial environmental justice: Government and governance in India. *Geoforum*, *37*, 660–670.

WWF International. (2012). *Shared risk and opportunity in water resources: Seeking a sustainable future for Lake Naivasha*. Switzerland: Author.

Xiaoming, Z. (2005). An analysis of Joseph Nye's "Soft Power" Concept [J]. *American Studies Quarterly*, *1*, 001. Retrieved from http://en.cnki.com.cn/Article_en/CJFDTotal-MGYJ 200501001.htm

Young, I. M. (1990). *Justice and the Politics of Difference*. Princeton, NJ: Princeton University Press.

Zeitoun, M. (2008). *Power and water in the Middle East: The hidden politics of the Palestinian-Israeli water conflict*. London and New York: IB Taurus.

Zeitoun, M. (2013). Global environmental justice and international transboundary waters: An initial exploration. *The Geographical Journal*, *179*(2), 141–149.

Zeitoun, M., & Allan, J. A. (2008). Applying hegemony and power theory to transboundary water analysis. *Water Policy*, *10*(2), 3–12.

Zeitoun, M., & McLaughin, K. (2013). Basin Justice: Using social justice to address gaps in river basin management. In T. Sikor (Ed.), *The justices and injustices of ecosystem services*. London and New York: Routledge.

Zeitoun, M., & Mirumachi, N. (2008). Transboundary water interaction I: Reconsidering conflict and cooperation. *International Environmental Agreements*, *8*, 297–316.

Zeitoun, M., Mirumachi, N., & Warner, J. (2011). Transboundary water interaction II: Soft power underlying conflict and cooperation. *International Environmental Agreements*, *11*, 159–178.

Zeitoun, M., & Warner, J. (2006). Hydro-hegemony – A framework for analysis of trans-boundary water conflicts. *Water Policy*, *8*(2), 435–460.

Zwarteveen, M. Z., & Boelens, R. (2014). Defining, researching and struggling for water justice: Some conceptual building blocks for research and action. *Water International*, *39*(2), 143–158.

Challenges in meeting water security and resilience*

Vijay P. Singh

ABSTRACT

This article addresses the following and related questions: (1) What is water security, and is it related to food security, energy security, health security and ecological security? (2) What factors impact water security? (3) What are the challenges in meeting water security? (4) What can be done to ensure water security? (5) Is water security entirely a technical problem? (6) What systems need to be engineered to help ensure water security? It further offers a personal perspective on water security and engineering solutions.

Introduction

Water is so fundamental in our lives that its role hardly requires discussion. Our civilizations developed along the banks of rivers or shores of lakes or near freshwater bodies. Without water, it would be virtually impossible to meet the food and fibre needs of the growing population. Our ecosystem would collapse were it not for water. We would not be able to produce enough energy if the needed water were not available. Our factories would not run in the absence of water. Our industries would not function if they did not have the water they need. Our transportation systems would come to standstill if water were not provided. Our economy would not grow if enough water did not exist, and we would not enjoy the economic prosperity that we do today. The list goes on. Suffice it to say that water and water security are a must. Yet, we have been less than prudent in managing this precious resource and the bodies that contain it. The result is that in many parts of the world the situation of water shortage has already reached crisis proportions. The question arises: Are we learning anything and doing something to make sure that we will have plenty of good-quality water in the years and decades ahead? Do our political systems promote long-term management of our water bodies?

The objective of this article therefore is to discuss water security and its relation to food security, energy security, health security and ecological security; factors impacting water security; the challenges that must be met for water security; and the systems that should be developed to ensure water security on a sustainable basis. The article also explores whether water security is entirely a technical problem or is much more than that, and concludes with a personal perspective on water security and engineering solutions.

---

*This article is based on the Ven Te Chow Memorial Lecture at the 16th World Water Congress in Cancun, Mexico, 1 June 2017.

Water security

Water security can be defined as access at all times to sufficient good-quality water to satisfy varied needs. Both quantity and quality, as well as access at all times, are the operative words here. Thus, water security is built on four pillars. (1) *Demand for and use of good-quality water* includes appropriate use based on the knowledge of quality water and treatment. (2) *Access to good-quality water* includes having sufficient resources and facilities to obtain adequate quantities of good-quality water to satisfy the need. (3) *Supply of good-quality water* includes the supply of sufficient quantities of good-quality water on a consistent basis. (4) *Availability* includes the identification and development of water resources. Availability decides the previous three pillars. The four pillars show that sufficient good-quality water must first be available. Then, it should be supplied, be easily accessible for use, and be used wisely.

Water security has space and time dimensions. The space dimension can range from an individual family unit to a village, town, district, state or province, country, continent, or the entire globe. Water security can vary from one spatial unit to another. For example, in a given state, certain parts are water-secure but other parts may be water-scarce. Likewise, the time dimension can vary from a day, week, month or season (winter, spring, summer or fall) to a year, decade, or century. A region can be water-secure in one part of the year but not in other parts. In parts of Rajasthan, in India, the lack of water becomes so acute in summer that thousands of people, along with their herds, leave their homes for northern parts of India to find water to survive. When the monsoon brings relief, they go back to their homes.

The time variability of water security is directly connected to the year-to-year variation in climate which impacts water availability and supply. For example, even a generally water-secure area can become water-scarce in years of prolonged drought. For convenience of analysis, water security can be classified into long-term, medium-term and short-term. This discussion shows that water security is two-dimensional and should be considered as such. An example of the two-dimensional characteristic is monsoonal countries in Asia. Consider India. During the monsoon season, when the north-eastern parts are deluged with floodwaters, certain north-western parts at the same time reel under a shortage of water. On the other hand, the same region may be hit by floods in one season and drought in another. Both the space and time dimensions of water security are affected by the need, demand, use, supply and availability of good-quality water as well as social, economic, political and cultural influences, which themselves may vary with space and time.

Demand for and use of water

People need water for a variety of uses, such as agriculture, livestock, fisheries, domestic consumption, industry, energy generation, waste disposal, recreation and mining.

Domestic water consumption

Each of us needs 15–20 litres of water per day to survive. The average American uses 378–666 L of water at home each day, whereas the average African uses about 20 L.

Besides drinking, water is needed for cooking, washing, bathing, sanitation, cooling and heating. Water consumption varies widely from country to country and from city to city. In general, the higher the standard of living, the greater the water use.

Kuwait has the highest per capita water consumption: it had 200 L of water consumption per person per day in the 1980s but it has gone up to 500 litres per person per day (lpd). The world is urbanizing rapidly. By 2050, 70% of the world's population will live in urban areas. In developing countries, cities grow by 5 million residents every month. Infrastructure does not keep up, resulting in expansion of slums and limiting access to water and sanitation. Mexico City, Mexico, has over 21 million people in the metropolitan area, and domestic water use is up to 364 lpd. Buenos Aires, Argentina, has over 12 million, using 378–400 lpd. Shanghai, China, has over 23 million, using 411 lpd. New Delhi, India, has over 20 million, using 78 lpd. This very high level of water consumption in many cities is not sustainable.

Water for food production

Agriculture is the largest consumer of freshwater, and globally its share is about 75% of all freshwater consumption. It has been estimated that nearly 1000 L of water is needed to produce 1 kg of grain, 15,500 L for 1 kg of beef, 20,000 L for 1 kg of coffee, 11,000 L for one quarter-pound hamburger, 7000 L for one T-shirt, 5000 L for 1 kg of cheese, 5000 L for 1 kg of rice, 3000 L for 1 kg of sugar, 2000 L for 1 L of milk, 1800 L for 1 kg of soybeans, 1000 L for 1 kg of wheat, and 900 L for 1 kg of corn (Hoekstra & Chapagain, 2008). Moreover, 5 L are needed to produce 1 L of bottled water, 30 L for one cup of tea, 140 L for one cup of coffee, 960 L for one loaf of bread, 50 L for an orange, 170 L for a glass of orange juice, 70 L for an apple, 190 L for a glass of apple juice, 2400 L for a dozen eggs, 3900 L for 1 kg of chicken meat, and 4800 L for 1 kg of pork. We often do not realize that it takes a lot of water to produce our daily diet.

Meaty American and European diets require 5000 L of water per day, whereas vegetarian African and Asian diets require 2000 L. Meat consumption in China has increased dramatically, from 20 kg per person per year in 1985 to 50 kg per person per year in 2009, which is equivalent to 390 km^3, the total amount of water used in Europe. Shifting dietary habits is almost impossible to reverse.

There is a need to reduce the agricultural share of water use, particularly in developing countries. This can be accomplished by improving irrigation efficiency and crop management, and developing species that require less water. A small reduction in the agricultural share of water use can go a long way towards ameliorating water shortage in other sectors. In developed countries, such as the United States, Canada, Australia and European countries, agricultural irrigation efficiency is much more efficient, but even there, efficiency could be increased. If irrigation efficiency in developing countries is not increased, the agricultural share will go up as population increases and more land is brought under irrigation. The question is, where will the extra water come from?

Industry is a major consumer of water in North America and Europe. For example, in Texas, USA, industry uses as much water as agriculture does. Comparing the total annual water withdrawal per sector by percentage, it is estimated that Asia uses 81% for agriculture, 11% for industry, and 7% for the domestic (including urban) sector; North

America, 39% for agriculture, 48% for industry, and 13% for domestic; Europe, 32% for agriculture, 53% for industry, and 15% for domestic; Latin America, 71% for agriculture, 10% for industry, and 19% for domestic; Africa, 86% for agriculture, 4% for industry, and 10% for domestic; Oceania, 73% for agriculture, 12% for industry, and 19% for domestic; and the Caribbean, 69% for agriculture, 8% for industry, and 23% for domestic.

Water for energy generation

Energy is used for water pumping, supply and conveyance, water and wastewater collection and treatment, water distribution and wastewater discharge. It is estimated that in the US electricity accounts for 75% of municipal water processing and distribution cost, and 4% of US power is used for water supply and treatment.

Electricity production is one of the largest users of water. On average about 25 gallons of water are used to produce 1 kWh. In the US, on average about 39% of freshwater withdrawals (excluding hydropower demand) is used for thermoelectric power plants (136 billion gallons or 500 billion L per day), 40% for irrigation, 13% for public supply, 5% for industrial use, 1% for domestic use, 1% for livestock, 1% for aquaculture, and 1% for mining. Electricity consumes 20% of water, excluding agriculture. For a 60-watt incandescent light bulb burning 12 hours a day for a year in 111 million houses, a power plant would consume about 655 billion gallons (2.5 km^3) of water. A study conducted at Virginia Tech (2008) investigated the water-use efficiency of different sources of energy generation. The study concluded that natural gas plants use 3 gallons of water per million BTU; hydroelectric power plants, 20 gallons; coal power plants, 41–464 gallons; liquid natural gas plants, 145 gallons; nuclear power plants, 2400–5600 gallons; fossil fuel thermoelectric power plants, 230–270 gallons; ethanol plants, 2500–29,100 gallons; biodiesel plants, 14,000–75,000 gallons; thus, biofuels are an irony when it comes to water. Currently 2% of irrigation water is used for energy crops. If all plans are implemented, 180 km^3/y of water will be needed, which is not sustainable.

In recent years hydraulic fracturing has gained momentum for developing oil resources. Natural gas production is projected to increase by 30% over the next 25 years. Over one million wells have already been hydraulically fractured since the 1940s. An estimated 35,000 wells are fractured every year. Hydraulic fracturing uses a mix of water, sand and chemical additives in 90% of wells in the US. It is a water-intensive process: 2.3–4.8 million gallons (8.7–18.2 million L) of fresh water is used per well. Once used, the water is extremely polluted and has to be treated before disposal. This raises serious environmental concerns. Fracking also causes groundwater contamination during drilling and production. Hence, proper treatment of wastewater must be provided for. Fracking also requires truck traffic that impacts water quality, can cause surface spills, and may require stormwater management. Further, water use for fracking may be competing with other uses.

Water and human health

About 80% of diseases in developing countries are due to lack of access to clean potable water. Pathogens transmitted through water kill 4 million people every year by amoeba-linked diarrhoea, cholera and typhoid, and about 3900 children die every day (World Health Organization, 2004). Ninety per cent of the 2.2 million deaths of children that occur each year are in children under 5 years of age. A child dies from a water-related disease every 15 seconds. Thirty per cent of water-related deaths are due to diarrhoea, and 84% of water-related deaths are in children under 15. About 98% of water-related deaths occur in the developing world. The most effective management intervention is to provide safe drinking water and proper disposal of human waste.

Availability of water resources

The total volume of water in the globe is about 1400 million km^3, 97–98% of which is saline, and only about 2–3% fresh (28–42 million km^3, average 35 km^3), unevenly distributed over and below the land. The total volume of freshwater, approximately 37.5 million km^3, that is available on the Earth is distributed as 29 million km^3 (about 77%) in polar ice caps and glaciers; 8.3 million km^3 (22%) in groundwater, about half at depths less than 800 m and half deeper than 800 m; 0.12 million km^3 (0.32%) in lakes, rivers and streams; 0.013 million km^3 (0.18%) in soil moisture and seepage; and 0.103 million km^3 (0.03%) as atmospheric water vapour. Thus, only a small percentage of freshwater is available in rivers, lakes and ponds, and much of this is highly polluted, further limiting freshwater availability.

The per capita availability of water depends on population and natural rainfall. The countries with the most abundant freshwater resources are Suriname, with 479,000 m^3, and Iceland, with 605,000 m^3. In terms of water per capita per year, Pakistan has 100–5000 m^3, and India 1000–1700 m^3; Egypt has the least, with 25 m^3, followed by the United Arab Emirates, with 61 m^3. In terms of per capita water in storage, North America has the most, with 6150 m^3. China has 2483 m^3, and India 130 m^3; Ethiopia has the least, with 43 m^3.

The rainwater that becomes available varies widely from one country or region to another and from one region to another in the same country. For example, in India certain parts of Rajasthan receive no more than 150 mm of rainfall a year, while certain north-eastern parts receive as much as 14,000 mm. Runoff also varies from place to place, and hence the same applies to flow variability. The land can be dry in one year but completely deluged in another year. The high rainfall variability poses a challenge for water resources management.

Current water situation

The current health of water bodies is not encouraging. An alarming number of rivers no longer reach the sea, including the Indus, Rio Grande, Colorado, Murray-Darling and Yellow, which are the arteries of main grain-growing areas. Freshwater fish populations are in precipitous decline. Fish stocks have fallen by 30% (World Wide Fund for Nature, 2015), the largest fall in population of animals in any ecosystem. Fifty per cent of the world's wetlands were drained, damaged or destroyed in the twentieth century. The volume of freshwater in rivers has fallen, causing an invasion of saltwater in deltas, changing the balance between freshwater and saltwater. There seems to be a global water crisis that is impacting supplies of food and generation of energy and other goods.

Consider Lake Chad. In 1963 it occupied areas in Chad, Niger, Nigeria and Cameron. By 2001 it had shrunk to a small fraction of its original size, occupying areas only in Chad and Cameron. Consider the Aral Sea, in Kazakhstan and Uzbekistan. Once the fourth-largest lake in the world, it has depleted to a new low, because water was diverted from the sea and its feeder rivers for decades to support cotton and rice cultivation in an extremely arid area. The eastern basin of the freshwater body has completely dried up. Drought and high evaporation have caused regional ecological, economic and health issues. Eighty-five per cent of its wetlands have been eliminated, and 50% of the birds and mammals have disappeared. Sand and salt are picked up by the wind and transported as far away as the Himalayas.

Lake Urmia in Iran has shrunk to 5% of its original volume in just 20 years. The Zayandeh River, one the largest rivers in the country, is now mostly dry. Lake Mead, the largest man-made reservoir in the United States, is the source of water supplies for millions of people in Arizona, Nevada and California. The lake was near full capacity in the summer of 1983 but had fallen to 37% by May 2016. Lake Powell, the second-largest water reservoir in the US, is likely to have its water level fall to dead pool, meaning it will no longer generate hydropower or flow through the gates. This alarming situation is due to drought and unsustainable use. Climate change can partly be blamed for long-term droughts. Similar examples abound all over the world.

Water shortage and crisis

Local water shortages are multiplying, and the amount of water being withdrawn is dangerously close to the limit, or over it. There have been severe and long droughts in Australia, Brazil, China, India, Russia, South America and the US. These droughts continue to occur more frequently. Already some years ago, many hydroelectric power plants in Brazil did not have enough water to drive turbines, and there were repeated brownouts.

Water stress is defined as a situation where there is less than 1700 m^3 of water per person per year, and *water scarcity* as less than 1000 m^3 per person per year. In 1995, 11 countries and a population of 270 million were water-stressed, and 18 countries and 166 million people were water-scarce. It is projected that by 2025, 15 countries and 2.3 billion people will be water-stressed and 9 countries and 1.7 billion people water-scarce. There will be absolute scarcity for 1 billion people by 2025, and many countries will import over 10% of cereals due to their inability to produce sufficient cereals for want of water.

Nearly 1 billion people, approximately one person in seven, do not have access to safe drinking water. This crisis is particularly severe in the more populous areas of Asia and Africa. By 2025, about two in three people will be living in water-stressed areas. The reasons for this situation are alarming rise in population, over-exploitation of groundwater, wasting of water, and mismanagement of water resources.

Many countries are facing a water crisis. In India, it has been building for decades and may soon reach a flash point (Biswas, Tortajada, & Saklan, 2017). The population grew from 1.03 billion in 2001 to 1.21 billion in 2011. In North India, groundwater depletion is a major concern; 80% of agriculture depends on it. The yearly average water availability per person in India was 1816 m^3 in 2001, but only 1545 m^3 in 2011, and may decline to 1340 m^3 by 2025. Nearly 54% of India faces high to extremely high water crisis. Fifty-four per cent of India's wells are depleting.

China has 30% of the world's population and only 7% of its freshwater resources. Along with its booming economy, China faces a looming water crisis. More than half of China's 50,000 rivers have disappeared. More than 400 cities have a water crisis, and 110 are facing serious water scarcity.

In Iran the water situation is dire. Daily consumption of water is 66 gallons per capita. It is the 24th most water-stressed country due to lack of freshwater and over-consumption. It is making use of 70% of renewed freshwater resources, far beyond the 40% acceptable limit. In South Africa, freshwater supply is under stress. By 2020 water requirements will double. In 2013, violent protests erupted in Protea South, Soweto, because people could not get enough water for sustenance. In 2014, in Mothutlung, police killed people who were protesting water scarcity.

Australia seems to be on the verge of drying out due to a shortage of rainfall, which may be due to climate change. Dams are now getting one-sixth of the normal water from rainfall. Desalination is being considered as an option to meet water needs. Europe is not immune to water crisis.

Factors affecting water security

The factors affecting water security are growing population, agricultural irrigation, rising domestic demand due to rising standard of living, increasing industrial demand, escalating energy consumption, mining, climate change, urbanization, deforestation, and migration of people.

Demography

Over the past 50 years the population has increased from 3 billion to 7.5 billion, but water use has tripled. Population is likely to rise by another 2 billion by 2025, and 3 billion by 2050. The demand for water will rise correspondingly, or even more. Habits and diets that make a big difference are changing to those that require more water. Assuming no change in agricultural efficiency, the world will need as much as 60% more water for agriculture, to feed 2 billion more mouths. This is roughly 1500 km^3 of water, as much as is used for all purposes in the world outside of Asia. The top seven water consumers in the world are India, using 13% of water resources, China 12%, United States 9%, Russia 4%, Indonesia 4%, Nigeria 3% and Brazil 3%; the rest of the world uses 52%.

Climate change

Weather is defined as the state of the atmosphere at a given time and place, with respect to variables such as temperature, moisture, pressure, wind, etc. *Climate* is the average weather, defined as the statistical mean conditions over a period of many years, typically two to three decades. *Climate change* is defined as changes in climate in excess of natural variability and attributable to human activity. It is caused by spiralling population, the high pace of industrialization, increasing use of fossil fuels in industry and transport, and deforestation for agriculture and urbanization. Its impacts encompass global warming, change in the hydrologic cycle, greater precipitation (and more uneven distribution), melting of glaciers and snow, sea-level rise, and greater frequency and intensity of extreme weather events. Climate change leads to uncertainty in water availability and supply, lower crop yields, higher demand for energy, loss of biodiversity, and higher health risks.

Climate change has implications for agriculture, food security, water security, energy production, extreme hazards, ecosystem health, rainforests, pollution, environment, migration of people, social well-being, ecosphere, biosphere and world peace. It is estimated that the temperature will rise by 2 °C over the next 100 years. The pattern of water use by plants will change, producing changes in how plants grow, reaction of trees to rain, drying up of biomass during droughts, and rapid growth of crops and then wilting.

Hydrological impacts of climate change include intensification of the hydrologic cycle; greater frequency of floods and droughts; greater risk of inland flash floods; coastal areas at risk due to more flooding; higher frequency of extreme events such as hurricanes and tornadoes; shrinking snowpack and changes in the timing of runoff from mountainous areas; glacial retreat in mountainous areas; coastal communities at risk from sea-level rise; and reduced freshwater availability and changes in water supply for human consumption. These changes will result in alterations in agricultural practices, lower yields from rain-fed agriculture and energy generation, more people exposed to water stress, and replacement of tropical forests by savannah.

Engineering water security

An integrated approach is fundamental to ensuring water security. The integrated approach must encompass Integrated Water Resources Management, integrated watershed management, and integrated water supply and use management. The integrated management must be multi-criteria, multi-objective, and multi-constraint; and must be the basis for decision making. Social, cultural, political, legal, environmental and economic considerations will constitute constraints within which the integrated management must be practised. It is vital that people, processes and policies are brought together in a seamless manner.

Integrated Water Resources Management requires that surface and subsurface waters are managed sustainably without harming the environment. Surface water resources management may entail management of rivers, streams, brooks, bayous, reservoirs, lakes, ponds and estuaries; and may require construction of hydraulic facilities for storage, diversion and transfer. These should be done with utmost care and with the full support of the stakeholders. Put succinctly, there must be a well-formulated surface water resources management policy or law that those who manage and those for whom the management is done understand the policy unambiguously. Groundwater management must be based on a well-formulated policy so that groundwater resources are not unduly exploited. Ideally, groundwater extraction and replenishment must be balanced, but this is not what is happening these days. Groundwater is being pumped as if the supply were unlimited. The result is that groundwater depletion has reached crisis proportions in many countries, and aquifers once emptied will not be revived.

Integrated Water Resources Management is practised at the river-basin or national scale.

Integrated watershed management requires that water resources management at the watershed scale is integrated with the entire ecosystem and the water resources of the watershed are managed in an integrated manner. The health of the watershed is the responsibility of the management and must be maintained so that it can be sustained for generations to come. This may entail conjunctive use of surface and groundwater, flood control, drought management, rainfall variability, runoff variability, floods and droughts, freshwater storage, pollution control, land use and land cover management, afforestation, land management and environmental management.

Integrated supply and demand management must consider water demand management, water requirements management, water use management, and water supply management. The integrated management must also consider the requirements and demands of different stakeholders, order of priorities, valuation or pricing of water, and social and economic considerations. More diverse portfolios for supply must be employed.

Since water is limited, it should not be wasted. Integrated management will have to simultaneously consider reducing water consumption, conservation of water, efficient use of water, treatment of water, reuse of water and recycling of water. Appropriate incentives must be provided for all these. Development of additional resources may be necessary where justified.

Water conservation must be practised. Every drop conserved is a drop that can be used for other purposes. Efficient irrigation must be developed. Irrigation methods must be improved so that every drop counts and does not go wasted. To that end, what is required is to determine crop water needs, crop area, crop type, soil type, climate, and irrigation methods and frequency. Tools for optimizing water use must be developed. Water treatment needs emphasis.

Water transfer from water-rich basins to water-deficient basins may be explored. For example, in the United States water is transferred from northern California to southern California, an arid and semi-arid region with a large population.

Planning for new projects must include climate change impacts and changing priorities. Existing infrastructure was designed using hydrologic records that did not reflect the effects of climate change. Therefore, a systematic revision of existing infrastructure is needed. For example, larger floods may overwhelm existing control structures. Reservoirs do not get enough water to store for people and plants during droughts. Snow and ice are natural regulators, storing water in winter and releasing it in summer, but they are now being disturbed. Global warming melts glaciers and causes what would have been snow to fall as rain instead. There will be more swings between flood and drought. In many areas, people are now willing to support construction of dams.

To develop a water, food and energy security system, partnerships among academia, the government sector, NGOs and the private sector (farmers) must be developed. The educational system that trains the trainers and teaches the teachers must be emphasized. Wasting food is a waste of water as well.

Key challenges

To ensure and sustain water security is one of the grand challenges of the twenty-first century. Water security is key to energy security, food security and ecosystem security. Sustainable development is not possible without water security. Fundamental to water security is proper water and water resources management, which involve technical as well as non-technical aspects. Technical issues are relatively easy to handle, because there is enough technical know-how to deal with most technical issues. It is the non-technical issues that are difficult to handle, especially in a democratic society. There are many conflicting interests and political considerations, so solutions acceptable to all parties are hard to come by.

Compounding the water security challenge is climate change. Although it is widely accepted that climate change is real and its causes are primarily human-induced, little is being done to arrest it. This is because of a number of reasons. First, climate has a global scale and it is not easy to have an international consensus on concrete measures and their implementation in a time-bound manner. Second, economic development and measures to control climate change are perceived to be not in alignment, especially in the near term. Third, democratic systems of government are not always conducive to making tough decisions and are more motivated by election cycles. Fourth, there are legal and environmental issues that have to be grappled with.

Social conflicts and wars are becoming a problem for development and for water security. These days conflicts are becoming only too frequent, and we are losing a sense of who we are and who we want to be. Imposition of social, political and religious beliefs is the root cause of many of the problems we are facing today.

Abatement of pollution of water bodies in developing countries is a huge challenge. Having laws and policies on the books is no better than not having them, if they cannot be implemented. This is one of the major problems in developing countries. Enforcing such laws is not easy. Waste management is another serious issue. In developing countries, littering is commonplace, and solid waste management can no longer be neglected if water security is to be achieved (Ahmad, 2017).

It is well known that water management for water security is as much non-technical as technical. Therefore, integration of engineering and technology with socio-economic and political sciences is urgently needed, but it is a challenge that we do not yet know how to cope with, or that we have not yet been able to successfully deal with. Organizations such as the International Water Resources Association that are dedicated to facilitating this sort of integration are few and far between.

Another major challenge is how to live in peace. We spend far too much money, manpower and resources on fighting each other, dominating each other, resolving conflicts and social ills, and creating problems in the first place, rather than developing an environment in which we all can live in peace, share resources so we can enjoy a high quality of life, help each other develop together, and truly grow as one family.

Re-examination and reflection

The question we must ask is: Where are we headed as a society? Our approach from development to management calls for a paradigm shift. We have achieved tremendous scientific and technological progress. However, it is not certain that the same can be said about social progress. Our values seem to be changing rapidly, and not necessarily for the better. We are witnessing enormous changes in the global demographic landscape. We are having too many conflicts and wars. We are constantly in a race of competition and consumption.

We often forget that fertile lands and resources are finite, and our standard of living does not have to be unsustainable. After all, the objective of life is far more than consumption, competition and comfort. Natural resources are God's gift. We must use them wisely. We owe it to future generations to leave them in better shape than we received them.

Conclusions

From this discussion the following conclusions are drawn.

(1) Water security requires management of water resources that are available, development of resources if feasible, management of surface and groundwater resources, management of water supply, management of water demand and requirement, and management of water use.
(2) The management of water use must include efficient use, reuse, recycling and treatment.
(3) Waste of water must be avoided.
(4) Pollution of water bodies must be curbed. This will also require waste management.
(5) There must be well-tested laws and policies for surface water resources and groundwater resources. These policies must be implemented like the law of the land.
(6) Integrated approaches to water management entailing technical and non-technical aspects must be employed.

Disclosure statement

No potential conflict of interest was reported by the author.

References

Ahmad, R. (2017). US water regulations and India's water challenges. *Journal AWWA, 109.3*, 65–67.

Biswas, A. K., Tortajada, C., & Saklan, U. (2017, March 15). *India is facing its worst water crisis in generations.* (pp. 1-3). *Quartz India.* Retrieved from https://qz.com/931878/india-is-facing-its-worst-water-crisis-in-generations

Hoekstra, A. Y., & Chapagain, A. K. (2008). *Globalization of water: Sharing the planet's freshwater resources.* Oxford: Blackwell Publishing.

World Health Organization. (2004). Water, sanitation and hygiene links to health: Facts and figures. Retrieved from www.who.int/water_sanitation_health/publications/factsfigures04/en

World Wide Fund for Nature (WWF). (2015). Crisis in global oceans as populations of marine species halve in size since 1970. Retrieved from https://www.wwf.org.uk/updates/crisis-global-oceans-populations-marine-species-halve-size-1970

The cultural dimensions of household water security: the case of Kathmandu's stone spout systems

Olivia Molden, Nicholas Griffin and Katie Meehan

ABSTRACT

This article contributes knowledge to the under-studied cultural aspects of household water security through the case of Kathmandu's ancient stone waterspouts. It asks why and how 'traditional' water supply systems persist as a form of water provision, and examines governance arrangements that pose challenges to these systems. It demonstrates that spout systems are critical sources of secure water supply, particularly for underserved populations. Also, the religious, cultural and social significance of spouts enables community autonomy and facilitates their persistence. However, conflicts between cultural heritage and drinking water law and policy undermine spout revitalization efforts and the entire system's integrity.

Introduction

Water supply involves a diverse array of delivery systems and governing institutions (Ahlers, Cleaver, Rusca, & Schwartz, 2014; Schwartz, Tutusaus Luque, Rusca, & Ahlers, 2015). Within these arrangements, many supply options – such as bottled water, tanker trucks, boreholes, sharing agreements and rainwater collection – are contemporary responses to alleviate water insecurity and public service failure at the household level (Bakker, 2010). Traditional water systems also continue to operate outside the public supply system, and play a role that goes beyond utilitarian purposes of securing reliable water supply (Lankford, Bakker, Zeitoun, & Conway, 2013). Bali's *subak*s (Lansing, 1987) and the *qanat*s of the Middle East and Central Asia (Bigas, Adeel, & Schuster, 2009) are well known and used primarily for agricultural purposes. Traditional urban systems, such as the ancient stone waterspouts of the Kathmandu Valley of Nepal, for example stone spout site Manga Hiti (Figure 1), are less well known. Yet, they are important sites of cultural organization and have provided water to city residents for approximately 1500 years – through multiple regimes and rulers, social crises, rapid urbanization and environmental disasters, such as the devastating earthquake of April 2015.

Drawing on the case of stone spouts in the Kathmandu Valley, this article asks: first, why and how have 'traditional' water supply systems *persisted* as a mode of water provision in modernizing urban areas; and second, what governance challenges

Figure 1. The spout site, Manga Hiti, pictured here, was quickly cleared of debris after the 2015 earthquake. Usually there are long lines of buckets waiting to be filled in the spout area. (Photo with permission from Karen Conniff.)

confront and constrain traditional systems? Based on a qualitative study of spout use and governance, we argue that traditional water systems are critical sources of water supply, especially for vulnerable, low-income residents. Beyond strictly utilitarian understandings of household water security, we also document their concomitant roles as sites of religious significance, cultural meaning and social organization – factors that we argue have enabled their persistence and reinvention within the service-provision configurations of Kathmandu. Despite such endurance, formal sector governance is testing the resilience of the spout systems. Drawing on analysis of institutional arrangements that currently govern spouts, we show that bifurcation between archaeological and water-provision mandates undermines efforts to revitalize the spout system as sources of water and cultural heritage.

This article contributes to scholarship that expands understanding of 'culture' in water management – not as a domain separate from politics, science and governance, but as implicated in the very meanings, mechanisms and modalities of managing water (Boelens, 2014). Cultural practices and traditions, defined here as the "beliefs, rituals, art and stories that create collective outlooks and behaviours" (Adger, Barnett, Brown, Marshall, & O'Brien, 2013, p. 112), directly inform how people know, use and manage the seemingly 'rational' world of water (Illich, 1985). With the rise of 'modern' water systems in cities, developed by Western science and engineering (Linton, 2014), traditional water systems have been dismissed by governments and decision makers as 'just cultural' and place-bound relics (Gibbs, 2010; Jackson, 2006; Maclean & The Bana Yarralji Bubu Inc., 2015), or as inefficient systems awaiting formalization and state control (Boelens & Seemann, 2014). Scholarship often portrays water security, for example, as rational, depoliticized decisions that follow universal economic behaviours and legal rules (Boelens, 2013; Boelens & Vos, 2012). This utilitarian focus neglects how the practice of securing water reveals place-based social networks, religious values, cultural meanings and asymmetries of power (Bakker, 2010; Mosse, 2008). Through the case of Kathmandu, this article extends these debates to demonstrate the vital cultural dimensions of securing and managing domestic water.

The article continues with a brief description of methods, materials and data interpretation. It then provides historical context for the degradation of the spouts system – a result of infrastructure modernization and the introduction of piped potable water in the 1950s. Drawing on qualitative and survey-based research, we document the multiple and overlapping roles of the spout systems. Spout systems are points of secure water supply for vulnerable groups, in that they provide free or inexpensive water sources. Spouts are also important spaces for social gathering, cultural identity and worship, particularly for Newar groups who originally built and managed much of the system. At the same time, we document how spouts occupy a regulatory blind spot, trapped in a patchwork of applicable laws governing cultural heritage and drinking water, which both cultivates their political autonomy and also undermines their resilience and longevity. As discussed in the final sections, this "forced engagement" (Boelens, 2013; Boelens & Seemann, 2014) between official and local systems reveals the urgent need to manage 'water security' in a way that extends beyond utilitarian concerns and basic livelihood strategies, to include vital cultural meanings and practices (Mosse, 2008).

Methods

The study focuses on the Lalitpur Sub-Metropolitan City, in the south-central part of the Kathmandu Valley. Lalitpur is Nepal's third-largest city and surrounds the ancient Patan Kingdom, a UNESCO World Heritage Site. Lalitpur is known for its rich cultural heritage and features a substantial number of functional spouts, making it ideal to examine the cultural resilience of spouts systems within urban agglomeration and modernization.

Materials for this study consist of a structured household survey (n = 351), semi-structured interviews (n = 41) and legal documents, all collected during three months of fieldwork in 2014. Survey data were provided by the International Centre for Integrated Mountain Development, who queried Lalitpur residents on domestic water sources, practices and use. Descriptive statistics and tests for correlation (Pearson's chi-squared and t-test), conducted by the lead author, provide information on household demographics, water access and source preferences, and perceptions of water stress. Qualitative data collection and interpretation, including a legal analysis of relevant laws, policies and regulations concerning stone spouts in the Kathmandu Valley, complimented survey results.

Compared to survey data, interviews with Lalitpur residents revealed important ethnographic insights regarding the religious significance, cultural values and community management of stone spouts. The lead authors conducted interviews with 11 household members, 10 community leaders, 3 spout historians, 9 international and non-governmental development experts and 8 government officials to gain a comprehensive view of the spouts. Because spout sites vary in physical condition and cultural significance, the authors recruited and interviewed leaders from nine different spout systems, as described in Table 1. To account for sociodemographic differences between migrant families (known as 'renters') and Newar homeowners (known as 'locals') in Lalitpur, the authors recruited a balance of such groups during interviews. Field notes were recorded during the interviews, and the interpretation of primary data was complemented by analysis of documents published by international organizations and the Nepalese government.

The history and degradation of the stone spouts

Stone spouts, known as *hiti* in Newari and *dhunge dhara* in Nepali, are carved water shrines and conduits, connected to larger cascading systems of ponds, canals, reservoirs and wells. The systems provide water for drinking, bathing, worship and a variety of other functions, depending on location and design (Table 1). Generally, a *hiti* filters and supplies water from interflow and base-flow processes, or through pipelines and canals that connect with upstream ponds, aquifers and springs (BeckerRitterspach, 1995). The spout system was Kathmandu's primary water source for at least 1500 years, until the mid-1900s, when metal pipelines were expanded for public use (M. Aryal, personal communication, 28 August 2014).

Newar people, who dominated cultural, political and economic life in the Kathmandu Valley until the Shah invasion in the mid-1700s, are accounted the original engineers and users of the spouts (BeckerRitterspach, 1995; Rademacher, 2011). For centuries, kings governed the entire system, while religious and familial trusts (*guthis*) took charge of local spout maintenance, rules and festivals (M. Aryal, personal communication, 28 August 2014). Between the 1950s and the 1970s, the introduction and expansion of the European-inspired water system led to a decline in spout use, as Kathmandu residents became more reliant on piped water and public standposts (Colopy, 2012). This shift reflected efforts to suppress Newar power during the Rana regime, but also to promote international development and capital investments in water supply to meet growing demands due to rapid urbanization (Rademacher, 2011; Shrestha, 2007). From the Rana coup in 1846 to the start of the civil war (1996–2006), the spout system received little formal attention (Regmi, 2005). Rapid urban growth and construction further destroyed many spouts and their water sources, as many groups – including government offices – treated spout, pond and aquifer sites as 'open space' rather than as common property and religious sites.

By the 1980s, the piped potable network fell into disrepair. Efforts to repair and extend the public grid continue at present, notably with the Asian Development Bank's formation of the Kathmandu Valley Water Supply Management Board and funding of the Melamchi project, a 26.3-kilometre water tunnel and treatment facility (Domènech, March, & Saurí, 2013). Progress is slow, and research speculates that once the project is

Table 1. Description of selected stone spout systems (NGOFUWS, 2009; Pradhananga, 2007).

Spout and place name (location)	No. of conduits	Period built	Cumulative flow (L/day)	Household users	Modifications	Direct uses
Sundhara (Urban; World Heritage Site)	4	Licchavi (400–750 CE)	199,543	500–1000	25,000 L underground tank, community building, standpost	Worship, drinking, bathing
Nag Bahal (Urban; World Heritage Site)	3	Licchavi	108,343	250	Wells, pipes, fountain, overhead tanks, distribution taps, community building; 40,000 L of spout water filtered and stored daily	Worship, washing
Nakabahil (Urban; World Heritage Site)	1	Destroyed 100 years ago by a Rana Prime Minister	Unknown	200	Well, standpost, and pump to overhead tank	Worship, drinking
Aalko (Urban)	5	Malla (1201–1769 CE)	683,105	225 (micro-grid), 500–600 (spout)	Rainwater recharge pits, ponds, overhead tank, underground pipes, community building	Worship, drinking, bathing
Washa Hiti and Amrit Hiti in Kumbeshwar (Urban)	2 in each	Malla	416,143	92 (micro-grid), 140–200 (spout)	Spout collection in 13,000 L underground tank, distribution taps, community building	Worship, drinking, bathing
Iku Hiti in Dhobighat (Urban)	3	Malla	480,715	200	Spout collection in underground tank, rainwater recharge, pumps, hoses, clothes washing areas, community pavilion	Worship, drinking, bathing, laundry
Harisiddhi (Peri-urban)	1 at several sites	Licchavi	Unknown	Unknown	Wells and ponds culturally more important than spouts	Worship, drinking, farming
Nau Dhara Hiti in Godavari (Rural)	9	Rana (1846–1951 CE)	Unknown	n/a	Repairs only	Worship, recreation

completed, the added flows from Melamchi will barely improve supply due to growth in water demand (Domènech et al., 2013; Muzzini & Aparicio, 2013). Against this backdrop, traditional water systems remain critical alternatives for city residents. As a news article reported, "By most international standards Kathmandu should be considered uninhabitable because the city of 2.5 million has few functioning water mains. But one of the reasons the city hasn't been abandoned is that its 400-year-old Malla-era water mains are still functioning" (Baidar, 2015).

Currently, spouts systems face several challenges. During the 1990s, many ran dry due to physical damage, disconnection and lowered water tables (Regmi, 2005). Today, an estimated 10% of the valley's 2.5 million residents rely on spouts as a water source (NGOFUWS, 2009), and they are a vital lifeline for households without piped water and for the millions of networked customers who experience frequent service interruptions and shortages (Shrestha & Shrestha, 2009). Spout water is now highly contaminated due to damage to ancient filtration systems and systemic failures in public sewerage and solid waste management (Warner, Levy, Harpp, & Farruggia, 2007). Additionally, access to more expensive sources often leads to the depletion of spouts downstream. Out of the Valley's 389 known spouts, 45 no longer exist and 68 have run dry, while 43 are 'illegally' connected to city supply lines and 233 still flow 'naturally' (NGOFUWS, 2009). Without water, many worry that spouts will become 'museum pieces' of urban modernity.

The cultural politics of stone spout systems

Spouts and water security

The greater Kathmandu metropolitan area can be characterized as water-insecure (Narain, Khan, Sada, Singh, & Prakash, 2013). Survey findings indicated that 75% of respondents felt their locality faced water crises, and 36% responded that their water needs were not regularly met. Although survey data indicated that 79% of respondents have access to piped potable supply, reliability is a major problem. Families must wait a week to receive 30 minutes of tap water, which is often supplied at an unannounced time. Quantities of water also vary: according to the municipal utility, approximately 154 million litres per day (MLD) are supplied in wet season and 95 MLD in dry season, when water demand is around 360 MLD (KUKL, 2010). Regardless of the season, water access is experienced unevenly – as caste, socio-economic status and geographic location matter greatly to the quantity, frequency and quality of water that families end up with (Narain et al., 2013; Warner et al., 2007).

In Lalipur, survey data indicated that 19% of residents use spouts. Analysis of survey responses found a significant relationship between household income, water source access and water stress; in other words, when households experience water stress, they tend to access spouts more often (Figure 2). When comparing spout-using households (n = 67) with non-spout-using households (80%, n = 284), spout-using respondents are typically of lower income, χ^2 (3, N = 351) = 10.016, p = .018, and fewer have access to networked services χ^2 (1, N = 351) = 10.259, p = .001. These findings confirm other studies that find spouts are important water sources for low-income families and/or those without connection to the grid, who thus must purchase water or rely on free

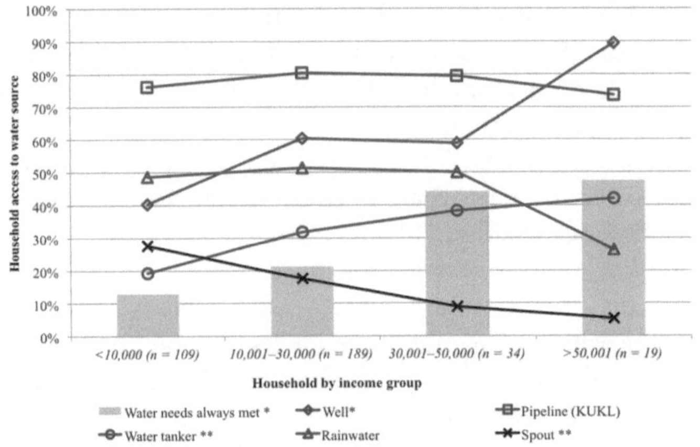

Figure 2. Household income in Nepalese Rupees, access to water sources and water needs met. (Income statistics consider the monthly total earned by all working members of the household. Pearson's chi-squared: $^{*}p < .05$, $^{**}p < .01$.)

public sources because they are squatting or renting (Muzzini & Aparicio, 2013; NGOFUWS, 2009; UN-HABITAT, 2008).

Interviews with spout users revealed the cultural complexities of securing water supply, including gender and caste dynamics. Nirmala, for example, moved to the outskirts of Lalitpur from a village in the mid-hills so that her children could go to school. In her rental home, which she shares with her two children and 12 other migrant families (each family rents one room in the house), residents have no access to piped water. With little money to spare, Nirmala expressed her worries and water stress. During the rainy season, Nirmala leaves out buckets to collect rainfall or draws milky water from the property's well, which she uses for cleaning purposes. In the dry season, her children have to wait several days to bathe or wear clean clothes. When time permits, she makes the 10-minute walk to Washa Hiti. There she collects one or two 10-litre buckets of water, which she strains through fine cloth to clean for cooking and drinking. During the dry season, she explained, she waits in line at the spout for an hour. When there is not enough water at the *hiti*, she buys water from the community reservoir or purchases jugs from a store. Nirmala expressed frustration that 'local' homeowners – typically Newars – have first rights to spout water. Despite these challenges, she prefers the spout because it is free, and of relatively trustworthy quality, with a strong local management system. Although she expressed no cultural attachment to the spout, she indicated that she respects local traditions and rules and finds religious value in the spout and its water. While imperfect models of community water supply, the spouts clearly provide a vital lifeline in an otherwise precarious urban waterscape.

Interview participants all want a safe, inexpensive, reliable and convenient water supply for their homes. Indeed, in the survey, 71% of respondents (n = 249) prefer municipal piped water because of its perceived convenience and quality. However, interviews also illustrated that many felt that such desires were merely wishful thinking due to the lack of progress on the Melamchi tunnel. After piped supply, spouts were the second preferred source of water. For several neighbourhoods, local management efforts and spout modifications have given residents easier access to water year-round. Survey findings suggest that spout-using respondents have a greater preference for community-managed water compared to respondents who do not use spouts (25% versus 16%, respectively), χ^2 (5, N = 351) = 12.961, p = .024.

Spout revitalization and social organization

In addition to their role as secure water supply, spouts are also important spaces for social organization, cultural identity, spiritual meaning and worship – factors that act as rallying points for social organization and action. Examples of social organization by residents are the 'spout committees' that have emerged in the past two decades to revitalize, protect and improve spout physical infrastructure and flows. Interviews with committee leaders indicate that spouts represent a sense of community autonomy and cultural resilience, given the political instability, poor government services and rapid changes in the urban landscape.

Spout committees are classic examples of collective action in water management. Committees meet regularly with residents, clean and maintain spouts, make repairs and modifications, enforce local rules and regulations, organize events and protests, and meet with relevant government agencies and non-governmental groups. Committees raise and manage funds from donations, membership fees, water use fees and loans from families, private parties or banks to pay for labour, repairs, modifications and festivals. Although all spout leaders interviewed were highly sceptical of the government – and several indicated zero trust in government involvement in spout activities – their testimonies revealed interactions and even collaborations with organizations, such as the Lalitpur municipality, local and international NGOs, and intergovernmental organizations (e.g. UN-HABITAT and UNESCO) to discuss spout issues, acquire additional funds and expertise for repairs and modifications, or work on public awareness campaigns. Committee leaders have also rallied spout users and neighbouring spout leaders to support court cases or protests in order to protect spouts, ponds and aquifers. Leaders and experts expressed worry that without greater government regulation and protection, spout systems will continue to degrade.

Community organization has resulted in examples of successful spout restoration and revitalization, as Table 1 documents. For example, at one spout, a Newar engineer initiated modern modifications in the early 1980s, by recommending that spout water be diverted into an underground reservoir at night to increase the storage potential. Leaders at Iku Hiti (Figure 3) followed some of the engineer's suggestions, and other neighbourhoods made similar changes, building underground and/or overhead reservoirs. Depending on the area, families can pay to access the stored water through distribution taps and standposts, or for some members, underground pipelines or hosepipes. For several neighbourhoods, modifications have also incorporated rainwater harvesting and recharge and building or revitalizing ponds and wells. Efforts to revitalize nearby ponds have also helped increase water flows to nearby spouts. For example, one community group and local NGO reclaimed an ancient *pokhari*, or pond, once used as a bus parking lot.

While repairs and modifications improve local water supply, such efforts also reflect shifting political arrangements facilitated, and motivated by, the cultural, spiritual and

Figure 3. Iku Hiti. Committee members are depicted cleaning the outflow pond, which will be used for washing laundry. The red walls surround the actual spout site. (Photo by lead author.)

social value of spouts. The very basis of spout organization is tied to cultural identity and ancestral ties, as indicated by the fact that all committee leaders were prominent Newar 'locals' in their communities. The cultural and ancestral meanings tied to spouts facilitate new social and political arrangements for certain spout neighbourhoods. For example, leaders and households of the Nag Bahal Hiti neighbourhood, located in the UNESCO Patan World Heritage Site, are all members of local Newar clans and rely on local water systems. This committee provides emergency and fire services, built a community health clinic, promotes local businesses and organizes women's, youth and elder groups. The Nag Bahal spout system functions like a micro-grid, as water is diverted from local spouts, wells and ponds into a chlorinated reservoir before it is distributed across the neighbourhood. According to interviews, households that use spouts indirectly (i.e. through pipes or hoses rather than direct collection) often have access to the piped potable network and can afford to buy water, yet they choose local water systems in an effort to preserve cultural ties and promote local self-sufficiency. In contrast, for the systems of Iku Hiti and Washa and Amrit Hiti – spouts in historically low-caste Newar neighbourhoods – the 'locals' interviewed explained that 'renters' were welcome and placed emphasis on the broader spiritual relevance of spouts, rather than Newar-specific traditions and festivals. Ultimately, all spout leaders explained that even if grid-based water supply became reliable, they were determined to provide their community with their own local water, even if they had to purchase tankers or build deep wells to fill the community reservoir.

In addition to the cultural and ancestral meanings attached to spouts, people continue to use, maintain and advocate for spouts as sites of worship. By providing 'pure' water for daily rituals and festivals, *hiti*s are important religious spaces for Hindus and Buddhists. The *makara* (mythological water guardians) carved into water conduits, frog and snake carvings, and idols and shrines in the vicinity are material symbols of the spiritually pure water that flows from spouts. In addition to daily worship and ritual cleansing, spout water is used in festivals such as Indra Jatra, Rato Machhindranath and Sithi Nakha, which focus on appeasing rain gods and appealing to water guardians. Historically, wealthy patrons built spouts for religious merit, to honour ancestors, or for the worship of Buddhist or Hindu water deities and guardians, such as the *naga*s (Gyawali, 2009; Sharma, 2003). For example, Washa Hiti is famous for its legendary medicinal properties, based on the story that a *naga* queen bestowed three spouts upon a tantric priest who healed her. The story of Washa Hiti and its proclaimed healing powers are so widely known that people from outside the community come to drink or bathe at the spout. Many spouts users expressed that if the same water came from a standard pipe rather than a stone spout, the water would not have the same religious and cultural significance. Thus, for community leaders and spout users, the cultural and religious significance of traditional infrastructure provides an understanding and source of water that extends far beyond utilitarian values or purposes.

A final reason that spouts persist relates to their social value as public spaces, something increasingly sought after in a rapidly developing city with little open space. The social value of spouts was clear from observations of people meeting and relaxing in spout areas, in addition to interviews with women who were often the primary water gatherers for their family. For example, one woman, who recently turned to buying tanker water to avoid the inconvenience of carrying water from the spout, indicated that she missed chatting with other women at the spout and was less informed about the community's social life. Despite complaints of long lines and heavy loads, all respondents indicated that they enjoyed going to the spout to meet and talk with other women in the neighbourhood to learn about the news, or upcoming festivals and weddings.

Challenges to spout governance

Despite the efforts of committees to revitalize and protect neighbourhood spouts, they increasingly run dry as the integrity of the entire system diminishes. Leaders and experts expressed frustration with the lack of government involvement, and explained that the government only valued spouts aesthetically. Such an approach runs counter to the widely held belief, by committee leaders and users, that spout water cannot be divorced from its cultural and spiritual value. Despite growing recognition and support from individuals in government departments, intergovernmental organizations, NGOs and corporate groups, higher-level decision making tends to disregard spouts. Gyawali and Dixit (2001, p. 703) have argued that the selection of development projects by governments and aid agencies is illustrative of "the 'cultural bias' of water management agencies and their inherent proclivities towards donor funded easy money and structural solutions to problems related to water". Similarly, several interviews suggested that proposals to integrate and revitalize the spout network were set aside because they were deemed not fundable by international donors. In addition to these issues, our investigation into the law and policies relevant to stone spouts noted that serious conflicts among different regulatory bodies contribute to a lack of meaningful action.

Stone spout management is primarily governed by four pieces of legislation: the 1956 Ancient Monument Preservation Act (Monument Act), the 1992 Water Resources Act (Water Act), the 2006 Water Supply Management Board Act (Management Board Act) and the 1999 Local Self-Governance Act (Self-Governance Act). The Monument Act preserves, and controls trade in, ancient monuments in Nepal. Under Article 2(a) of the Monument Act, stone spouts are classified as ancient public monuments if they (1) are more than 100 years old, (2) are important to history, arts, or science, and (3) have "specific value from a national or international point of view". Per Article 3(B) of the Monument Act, ownership of all public ancient monuments in Nepal vests in the Department of Archaeology. Yet, while the act requires the Department of Archaeology to preserve stone spouts to "keep the monument in its original form", the act does not protect a stone spout's function as a water tap and and site for worship and ritual. The Monument Act only ensures preservation of the stone sculpture and surrounding architecture. Thus, stone spout users can request aid from the Department of Archaeology to restore the physical aspects of ancient spouts, but not to restore water flow to spouts that have run dry.

The Water Act is umbrella legislation governing all water resource management in Nepal. The Water Act serves to consolidate "management, conservation, and development of water resources" under state authority (Water Resources Act, 1992). Applied to stone spouts, Article 3 of the Water Act provides that all water, including stone spout water, is property of the state. However, Article 4 establishes that stone spout water users have first priority or right of access, because spout water is primarily for "drinking and other domestic use". Operating as Nepal's procedural framework for water management, the Water Act does not designate any enforcement agency. Instead, later acts such as the Management Board and Self-Governance Acts delegate water management responsibilities.

The Management Board Act vests ownership of all water supply and sewage system assets in the Kathmandu Valley Water Supply Management Board (KVWSMB). Article 2(h)–(i) provides that KVWSMB is responsible for operating the water grid as a "water supply service" that consists of "the provision of water to … users upon processing and purifying the natural water". The scope of natural water accessible to KVWSMB includes "water coming out from stone spouts" in addition to more common sources associated with the water grid, such as surface and ground waters, lakes, rivers, streams and ponds. Thus, KVWSMB has the authority to appropriate stone spout water as a resource for public use. Interestingly, the Management Board Act does not consider stone spouts part of the water grid, but only refers to the water flowing from stone spouts as an accessible resource.

The Self-Governance Act is a broad measure towards the decentralization of governing powers to local bodies, such as a Village Development Committee, municipality, or District Development Committee. Article 96(c)(1) delegates water management responsibilities to these local bodies, giving them authority to "preserve rivers, streams, ponds, deep water, wells, lakes, stone water taps etc. and utilize, or cause [such waters] to be utilized". Notably, where the Management Board Act grants the water grid access to stone spout water for the purposes of public water provision, the Self-Governance Act grants local bodies authority over stone spout water use and preservation.

By occupying a convergence between cultural heritage and drinking water, laws governing stone spout management have splintered across the Monument, Water, Management Board and Self-Governance Acts over time. Consequently, rather than foster a surplus of stone spout conservation measures, the patchwork of applicable law collectively places spouts in a regulatory blind spot. This blind spot stems from the lack of clarity as to the role that each regulating body should assume in relation to other bodies' respective mandates. This legal ambiguity fosters an observable dynamic between agencies and community-level stone spout managers. It also demonstrates how failing to account for the cultural politics of water provision leads to a conflicting regulatory framework governing stone spouts.

Namely, the Self-Governance Act deregulates stone spout governance to accommodate community-level decision making over stone spouts. Governing local departments, the Department of Archaeology and KVWSMB, adopt the perspective that communities should assume financial responsibility as well. Thus, tension between communities and government agencies often arises in the context of stone spouts, as communities lack the resources to actualize management plans (Pradhan & Meinzen-Dick, 2003). The resulting effects for spouts are the perpetuation of conflicting conservation strategies, improper burden shifting, and complacency in assuming that another regulating body has taken up management efforts. Also, issues of corruption, funding, bureaucracy and overarching political instability contribute to the overarching neglect of stone spout conservation. Despite these issues, many government officials expressed that they respect the spout system and wish they could do more to assist community efforts to protect it.

Discussion

This article argues that 'traditional' stone spout water supply systems persist as a contemporary system of urban water provision, and not a temporary alternative. This is because spouts provide vital water supply, especially for vulnerable, low-income residents, but also because of their associated religious significance, historical meaning and social values. In a city with failing government services and rapidly changing cultural and physical landscapes, spouts provide added security and a sense of community, particularly for historic Newar residents. Community organizing around these cultural and functional dimensions has facilitated spout revitalization and repurposing to enable greater local self-sufficiency and political autonomy, despite incursion and increased control of water infrastructure by government authorities (Boelens, 2014; Meehan, 2014). Diverse cultural practices and meanings attributed to spouts and their water contribute to the persistence and transformation of the ancient system as a modality of water provision.

However, efforts to govern and preserve spout systems have always faced challenges. Attempts to create a modern Nepali state under the Rana regime's directives left the spout system in substantial disrepair, and many spouts were intentionally destroyed to make room for new construction and infrastructure. Following the public expansion of the European-inspired pipeline system in the 1950s, political priority focused on expanding the grid system for public water provision. In this context, many spouts' aesthetic and archaeological qualities were of little concern to the government until the 1990s, with the rise of spout committees and community protests. Frustrated with government services and politics, community committees and experts working in NGOs, corporations and government departments have started to advocate for spouts as cultural heritage and sustainable water sources. While these efforts can privilege 'local' users and a specific vision of Kathmandu's cultural heritage, community-level revitalization has strengthened decentralized water systems that provide a relatively inexpensive and convenient water supply. Yet, these efforts are limited in scope, as communities lack the ability to regulate the use of upstream water resources.

In particular, community-level management is impeded by conflicts between the jurisdiction of different acts and departments pertinent to spouts. Policy analysis revealed that bifurcation between archaeological and water-provision mandates leaves communities in a regulatory blind spot, and unable to receive adequate support to ensure water flows. The separate and differential treatment of spouts – as types of cultural heritage or public water – reflects development trends elsewhere in the world, where "setting up separate compartments or categories and defining and adopting key terms, such as cultural values, represents a means of exercising power over 'rights to define what is and not valuable about environments'" (Jackson, 2006, p. 26). By neglecting spouts, development groups and governments reinforce universal standards of urban water security, which focus on management or engineering strategies to improve the quality, quantity, reach and reliability of modern grid systems (Furlong, 2014; Joyeeta & Van der Zaag, 2009; Loftus, 2015; Meehan, 2014). As a result, diverse knowledge systems, identities and beliefs that influence why and how people use and manage traditional infrastructure remain largely overlooked in water resource frameworks (Pradhan & Meinzen-Dick, 2003). For Kathmandu, this "forced engagement" (Boelens, 2013; Boelens & Seemann, 2014) of official and local systems has resulted in greater water insecurity and further marginalization of disadvantaged groups and cultural heritage practices.

While many argue that water security is a political process (Bakker, 2010; Boelens, 2013; Lankford et al., 2013; Loftus, 2015), this study demonstrates that water security is also a *cultural* process, replete with religious and cultural significance. These findings extend Mosse's (2008, p. 943) assertion that "water systems have to be understood in their cultural specificity . . . : firstly, to reveal the wider set of social networks, values and meanings that water brings; and second, to see the part that water itself plays in the cultural specifics of power", in ways that underscore how household water security and management practices are in fact informed by 'more-than-utilitarian' dynamics and tensions of water supply.

What would recognition of cultural resilience and traditional infrastructures imply for improved water management in Kathmandu? Formalization of traditional systems is an inherently political process, and may serve to weaken – rather than strengthen – local water security (Boelens & Seemann, 2014) or introduce contradictory legal mandates and power dynamics (Meehan & Moore, 2014). Recognizing the cultural value of water in Kathmandu would require more than formal recognition of spouts as legitimate water sources and forms of tangible and intangible cultural heritage. As many leaders interviewed for this study argued, government action and development efforts should help ensure *both* piped municipal water *and* public water flows beyond the grid, for example by preventing over-extraction of spout tanks and promoting rainwater harvesting alongside restoration of the ancient water system. Ensuring spout flows and cultural heritage safeguards together provides people with choices that matter to their cultural identities and religious beliefs, as well as their daily needs.

Conclusion

Research presented in this article documents the contemporary ways people use and govern ancient stone spouts in Kathmandu. Findings indicate that cultural identities and beliefs are central to how and why people access or organize around spouts and ultimately prevent their erasure in a rapidly changing landscape. As in many other places where 'traditional' or indigenous water management systems are seen as 'just cultural' categories, or where ancient water systems are presented as historical relics, Kathmandu's spouts and their users remain peripheral to modern urban development efforts and standards. New pipelines and wells physically cut off spouts from their sources, and contemporary legal institutions keep spouts in a regulatory blind spot. Despite these challenges, spout systems represent a form of cultural resilience, where people interact with water infrastructure in ways that extend beyond utilitarian concerns.

Government involvement is key to ensuring equitable and long-term water supply for Kathmandu residents, as spouts are imperfect forms of household water security, especially given their concomitant roles as cultural, ancestral and religious sites. Yet, decision makers lack the ability to protect and revitalize both the cultural and the hydrologic integrity of spouts, due to conflicts between archaeology, local governance, and water supply mandates. Without an appropriate level of government support and intervention, and without meaningful involvement of spout committees in the realm of water management, rapid rates of urbanization and water extraction threaten the cultural resiliency of Kathmandu's celebrated and unique water infrastructure.

Acknowledgements

We are grateful to Dr Aditi Mukherji, Dr Daniel Buck, Abby Snyder, Gyanu Maharjan, Raju Maharjan, Govinda Joshi, Basantha Shrestha, Prof. Mukunda Aryal, the International Centre for Integrated Mountain Development survey team (Valerie Gamao, Manzari Singh, Mira Khadka, Nabina Lamichanne and Devjit Roy Chowdhury) and all the research participants who made this work possible. We also sincerely thank the editors of *Water International* and an anonymous reviewer.

Funding

University of Oregon Center for Asian and Pacific Studies Small Professional Grant; Association of Pacific Geographers Larry Ford Field work Scholarship in Cultural Geography and Margaret Trussell Scholarship; Wake Forest University School of Law Hopkins Pro Humanitate Public Interest Law Organization Grant.

References

Adger, W. N., Barnett, J., Brown, K., Marshall, N., & O'Brien, K. (2013). Cultural dimensions of climate change impacts and adaptation. *Nature Climate Change*, *3*(2), 112–117. doi:10.1038/nclimate1666

Ahlers, R., Cleaver, F., Rusca, M., & Schwartz, K. (2014). Informal space in the urban waterscape: Disaggregation and co-production of water services. *Water Alternatives*, *7*(1), 1–14.

Ancient Monument Preservation Act, Kathmandu, Nepal (1956).

Baidar, U. (2015, March 20). Going with the flow. *Nepali Times*. Retrieved from http://nepalitimes.com/article/nation/Going-with-the-flow,2116.e

Bakker, K. (2010). *Privatizing water: Governance failure and the world's urban water crisis*. Ithaca, NY: Cornell University Press.

BeckerRitterspach, R. (1995). *Water conduits in the Kathmandu Valley*. New Delhi: Munshiram Manoharlal.

Bigas, H., Adeel, Z., & Schuster, B. (2009). *The United Nations world water development report 3: Water in a changing world*. United Nations Educational, Scientific and Cultural Organization. Retrieved from http://www.unesco.org/new/en/natural-sciences/environment/water/wwap/wwdr/wwdr3-2009/

Boelens, R. (2013). The shotgun marriage: Water security, cultural politics, and forced engagements between official and local rights frameworks. In B. Lankford, K. Bakker, M. Zeitoun, & D. Conway (Eds.), *Water security: Principles, perspectives and practices* (pp. 239–255). New York, NY: Routledge.

Boelens, R. (2014). Cultural politics and the hydrosocial cycle: Water, power and identity in the Andean highlands. *Geoforum*, *57*(1), 234–247. doi:10.1016/j.geoforum.2013.02.008

Boelens, R., & Seemann, M. (2014). Forced engagements: Water security and local rights formalization in Yanque, Colca Valley, Peru. *Human Organization*, *73*(1), 1–12. doi:10.17730/humo.73.1.d44776822845k515

Boelens, R., & Vos, J. (2012). The danger of naturalizing water policy concepts: Water productivity and efficiency discourses from field irrigation to virtual water trade. *Agricultural Water Management*, *108*, 16–26. doi:10.1016/j.agwat.2011.06.013

Colopy, C. (2012). *Dirty, sacred rivers: Confronting South Asia's water crisis*. New York, NY: Oxford University Press.

Department of Archaeology. (2011). Kathmandu, Nepal. Retrieved from http://www.doa.gov.np

Domènech, L., March, H., & Saurí, D. (2013). Contesting large-scale water supply projects at both ends of the pipe in Kathmandu and Melamchi Valleys, Nepal. *Geoforum*, *47*, 22–31. doi:10.1016/j.geoforum.2013.02.002

Furlong, K. (2014). STS beyond the "modern infrastructure ideal": Extending theory by engaging with infrastructure challenges in the South. *Technology in Society*, *38*, 139–147.

Gibbs, L. M. (2010). "A beautiful soaking rain": Environmental value and water beyond Eurocentrism. *Environment and Planning D: Society and Space*, *28*(2), 363–378. doi:10.1068/d9207

Gyawali, D. (2009). Water and conflict: Whose ethics is to prevail? In M. Llamas, R. Martinez-Cortina, & A. Mukherji (Eds.), *Water ethics*. Boca Raton, FL: CRC Press.

Gyawali, D., & Dixit, A. (2001). Water and science: Hydrological uncertainties, developmental aspirations and uningrained scientific culture. *Futures*, *33*(8–9), 689–708. doi:10.1016/S0016-3287(01)00014-3

Illich, I. (1985). *H_2O and the waters of forgetfulness: Reflections on the historicity of "stuff"*. Berkeley, CA: Heyday Books.

Jackson, S. (2006). Compartmentalising culture: The articulation and consideration of indigenous values in water resource management. *Australian Geographer*, *37*(1), 19–31. doi:10.1080/00049180500511947

Joyeeta, G., & Van der Zaag, P. (2009). The politics of water science: On unresolved water problems and biased research agendas. *Global Environmental Politics*, *9*(2), 14–23. doi:10.1162/glep.2009.9.2.14

KUKL. (2010). *Kathmandu Upatyaka Khanepani Limited sixth anniversary report*. Kathmandu, Nepal: KUKL.

Lankford, B., Bakker, K., Zeitoun, M., & Conway, D. (Eds.). (2013). *Water security: Principles, perspectives, and practices*. New York, NY: Routledge.
Lansing, J. S. (1987). Balinese "Water Temples" and the management of irrigation. *American Anthropologist, 89*(2), 326–341. doi:10.1525/aa.1987.89.issue-2
Linton, J. (2014). Modern water and its discontents: A history of hydrosocial renewal. *Wiley Interdisciplinary Reviews: Water, 1*(1), 111–120. doi:10.1002/wat2.1009
Local Self-Governance Act, Kathmandu, Nepal (1999).
Loftus, A. (2015). Water (in)security: Securing the right to water. *The Geographical Journal, 181* (4), 350–356. doi:10.1111/geoj.2015.181.issue-4
Maclean, K., & The Bana Yarralji Bubu Inc. (2015). Crossing cultural boundaries: Integrating indigenous water knowledge into water governance through co-research in the Queensland Wet Tropics, Australia. *Geoforum, 59*, 142–152. doi:10.1016/j.geoforum.2014.12.008
Meehan, K. M. (2014). Tool-power: Water infrastructure as wellsprings of state power. *Geoforum, 57*, 215–224. doi:10.1016/j.geoforum.2013.08.005
Meehan, K. M., & Moore, A. W. (2014). Downspout politics, upstream conflict: Formalizing rainwater harvesting in the United States. *Water International, 39*(4), 417–430. doi:10.1080/02508060.2014.921849
Mosse, D. (2008). Epilogue: The cultural politics of water – A comparative perspective. *Journal of Southern African Studies, 34*(4), 939–948. doi:10.1080/03057070802456847
Muzzini, E., & Aparicio, G. (2013). *Urban growth and spatial transition in Nepal: An initial assessment*. World Bank. Retrieved from http://issuu.com/world.bank.publications/docs/9780821396599/88
Narain, V., Khan, M. S. A., Sada, R., Singh, S., & Prakash, A. (2013). Urbanization, peri-urban water (in)security and human well-being: A perspective from four South Asian cities. *Water International, 38*(7), 930–940. doi:10.1080/02508060.2013.851930
NGOFUWS. (2009). *Traditional stone spouts: Enumeration and mapping*. Kathmandu: NGO Forum for Urban Water & Sanitation.
Pradhan, R., & Meinzen-Dick, R. (2003). Which rights are rights? Water rights, culture, and underlying values. *WaterNepal: Journal of Water Resources Development, 9*(10), 37–61.
Pradhananga, S. (2007). *The stone spouts of Lalitpur Sub-Metropolitan City*. Kathmandu: Lalitpur Sub-Metropolitan City Press.
Rademacher, A. (2011). *Reigning the river: Urban ecologies and political transformation in Kathmandu (New ecologies for the twenty-first century)*. Durham: Duke University Press.
Regmi, A. (2005). Complexities of water governance: The rise and fall of groundwater for urban use. In D. Roth (Ed.), *Liquid relations contested water rights and legal complexity*. New Brunswick, NJ: Rutgers University Press.
Schwartz, K., Tutusaus Luque, M., Rusca, M., & Ahlers, R. (2015). (In)formality: The meshwork of water service provisioning. *Wiley Interdisciplinary Reviews: Water, 2*(1), 31–36. doi:10.1002/wat2.2015.2.issue-1
Sharma, S. (2003). Water in Hinduism: Continuities and disjunctures between scriptural canons and local traditions in Nepal. *WaterNepal: Journal of Water Resources Development, 9*(10), 215–247.
Shrestha, B. G. (2007). Chapter 8. Ethnic nationalism in Nepal and the Newars 2007. In M. Lawoti (Ed.), *Contentious politics and democratization in Nepal* (pp. 199–225). New Delhi: Sage .
Shrestha, S., & Shrestha, B. K. (2009). The influence of water in shaping culture and modernization of the Kathmandu Valley. In J. Feyen, K. Shannon, & M. Neville (Eds.), *Water and urban development paradigms: Towards an integration of engineering, design and management approaches* (pp. 183–188). Boca Raton, FL: CRC Press.
UN-HABITAT. (2008). *Water movements in Patan with reference to traditional stone spouts*. Kathmandu: UN-HABITAT Water for Asian Cities Programme Nepal.
Warner, N. R., Levy, J., Harpp, K., & Farruggia, F. (2007). Drinking water quality in Nepal's Kathmandu Valley: A survey and assessment of selected controlling site characteristics. *Hydrogeology Journal, 16*(2), 321–334. doi:10.1007/s10040-007-0238-1
Water Resources Act, Kathmandu, Nepal (1992).
Water Supply Management Board Act, Kathmandu, Nepal (2006).

Contested access: improving water security through benefit sharing

Bimo A. Nkhata

ABSTRACT

This article examines the potential and limits of benefit-sharing regimes in improving water security. It draws on a case study of benefit-sharing regimes focusing on the non-consumptive use of the Zambezi River basin. It demonstrates how these regimes have transformed over time from hierarchical to egalitarian benefit-sharing arrangements through the application of conflict-resolution mechanisms. This case analysis is aimed at informing benefit-sharing policies for African river basins.

Introduction

River basins are known to provide a range of benefits that buttress water security (Petersen-Perlman, Veilleux, Zentner, & Wolf, 2012). These benefits are varied, but generally include multifaceted direct and indirect goods and services such as domestic water, fish stocks, biodiversity, irrigation, power generation, transport, recreation, and scenic value. In southern Africa, improving water security requires an emphasis on securing access to such benefits (Turton, 2008). In particular, issues surrounding the sharing of benefits derived from river basins raise significant and interesting questions about the nature and substance of water security. While many of the benefits of river basins are readily apparent, the role of benefit-sharing regimes in ensuring effective water security usually goes largely unrecognized. Yet, the relationship between water security and benefit sharing is one of mutual significance.

This article explores the potential and limits of benefit-sharing regimes in improving water security in one of Africa's river basins, the Zambezi. I draw on a case study of the non-consumptive use of the river system to illustrate the evolution of benefit-sharing regimes in the context of local water security. The case study focuses on a section of the river basin (Lower Zambezi National Park) to demonstrate how over time these regimes transformed from hierarchical to egalitarian benefit-sharing arrangements. The study shows how initially a government agency erroneously assumed that access to river-basin-related benefits is uniform, stable and premised on a small number of preferences. This made it difficult for the government agency to acknowledge the dynamics and variability of access options as framed by different water user groups. This study is particularly significant due to its focus on non-consumptive use of water, a subject that is rarely addressed in water-security-related studies. It is aimed at informing benefit-sharing policies for improved water security in African river basins. It is envisaged that the study will contribute to societal efforts aimed at designing the next generation of benefit-sharing models for effective water security.

Conceptualizing water security and benefit sharing

What is water security?

The term water security means different things to different people, and different authors use different ways to define it. The Global Water Partnership (2000) considers water security as an overarching goal whereby all people have access to adequate, safe and affordable water. Grey and Sadoff (2007, p. 548) define water security as the 'availability of an acceptable quantity and quality of water for health, livelihoods, ecosystems and production, coupled with an acceptable level of water-related risks to people, environments and economies'. Petersen-Perlman et al. (2012) simplify this to the availability of freshwater in the right quantity and quality. The UN-Water Task Force on Water Security views the term as denoting the capacity of a population to safeguard sustainable access to adequate quantities of acceptable quality water (United Nations University/Institute for Water, Environment & Health [UNU/INWEH], 2013). Although somewhat simplified, these definitions represent diverse approaches for framing and promoting effective water security. However, it is now widely accepted that the success of particular water security initiatives in large part depends on how well they ultimately correspond to local demands, experiences and expectations (Hoekstra, Buurman, & van Ginkel, 2018).

While water security in general is dependent upon secure access to water of appropriate quantity and quality (Grey & Sadoff, 2007), how the concept is framed and applied does not adequately portray the ways in which water is allocated to meet the challenges of a changing world (Chomba, Hill, Nkhata, & Förster, 2017). In particular, the way water is allocated among multiple, ever-changing societal demands does not usually acknowledge the role of water security in transforming societies so that they are better able to prepare for and direct change (Pahl-Wostl, Lebel, Knieper, & Nikitina, 2012). The concept of water security is more often than not narrowly associated with efforts aimed at merely maximizing availability and minimizing risks, as if how water-allocation decisions are made is not important. In the face of this underestimation, concerns regarding the execution of local water security initiatives are emerging that suggest a need to pay attention to water-allocation processes (Fischhendler, 2013; Mekonnen, 2010; Zeitoun, 2011).

Although access to water has transformed societies over thousands of years (Angelakis, Mays, Koutsoyiannis, & Mamassis, 2012), as relative scarcity increases, we cannot simply continue merely trying to maximize availability and minimize risks. We need to expand our framing and application of water security. We must become more conscious of the role water-allocation processes and decisions can play in directing change and societal transformations (Chomba et al., 2017). In southern Africa, in particular, benefit-sharing regimes are required as we seek optimal allocations to secure the gains and to manage the losses that underpin water security efforts. To appreciate the policy relevance of benefit sharing in southern Africa, one needs to deeply reflect on what is currently happening in societies that are heavily and directly dependent on water resources for socio-economic development and poverty alleviation. Not surprisingly, perhaps, societies with relatively abundant water resources have had higher incidence of conflict and have tended to suffer from poor water governance, including benefit-sharing processes (Nkhata, Breen, & Mosimane, 2012a).

Based on the foregoing and building on the relevant literature, in this article 'water security' broadly refers to the capacity of human actors to safeguard equitable sharing of the benefits derived from river systems. It is important to note that this characterization of water security is not meant to be exhaustive; rather it is designed to be illustrative. While a broader characterization could perhaps have provided additional valuable insights, the case study presented in this article is still instructive particularly in illustrating why and how benefit-sharing regimes evolve over time to influence the substance of water security. This is because water security in river basins occurs under contested and uncertain conditions (Nkhata et al., 2012a). Decisions regarding who gets access to river-basin-related benefits and how associated rights to access are negotiated always involve complex value judgements. Choices about access options frequently lead to decisions that are a source of controversy and conflict. Participants in benefit-sharing regimes usually find it difficult to justify decisions regarding choices that go awry. The use of appropriate benefit-sharing mechanisms is extremely important in such complex situations, not only so that decisions can be made in a purposeful and consistent manner, but also so that decisions can be defended (Boyle, Kay, & Pond, 2001). Benefit-sharing arrangements are complex, interlinked social and hydrological processes that facilitate the allocation of gains from water security to a range of designated participants (Nkhata et al., 2012a). Whereas the production and flow of benefits from river basins result from both social and hydrological processes, the ways in which the gains from water security accumulate to beneficiaries are governed by complex societal norms, rules and organizational processes (Turton, 2008).

What are benefit-sharing regimes?

It was not until 1992 that the Convention on Biological Diversity (CBD, 1992) formalized the concept of benefit sharing in international environmental law and governance. Essentially, the concept denotes forms of social accountability and responsibility to direct returns from the use of natural resources, be they consumptive or non-consumptive, back to a range of designated participants within socially designed arrangements (Hayden, 2003). Various benefit-sharing regimes are increasingly being promoted as a means of addressing problems related to water governance, particularly in developing countries (Nkhata et al., 2012a; Turton, 2008). Such regimes seek to address fundamental questions about water justice, human behaviour, and sustainability. In this context, the term 'regime' refers to a body of fundamental rules and rights that systematizes society and ultimately influences the sharing of benefits (Vatn, 2010). As Ostrom (2005) illustrates, these rules and rights exist in diverse social settings such as in the home, in the neighbourhood, in local, regional, national and international councils, and in governments, firms and markets. A benefit-sharing regime, therefore, provides a set of conditions and a normative framework that directs the water-allocation decisions and actions of managers and users of water-based resources.

Table 1. Three types of benefit-sharing regimes for exploring water security settings.

Type of benefit-sharing regime	Governance arrangements	Examples
Hierarchical	Centred on command-and-control policy instruments established to support a central agenda	Government-led water management schemes, and catchment management authorities
Distributive	Involve voluntary exchanges established to support *quid pro quo* relationships	Payments-for-ecosystem-services initiatives, clean development mechanisms, and voluntary carbon market projects
Egalitarian	Based on the principle that all people are equal and deserve equal rights and opportunities	Access and benefit sharing initiatives, and community-led water management schemes

Source: Adapted from Nkhata, Mosimane, Downsborough, Breen, and Roux (2012b).

In this article, three types of benefit-sharing regimes provide a useful basis for exploring water security settings: hierarchical, egalitarian and distributive (Nkhata et al., 2012a) (Table 1). Hierarchical regimes involve benefit-sharing arrangements centred on command-and-control policy instruments established to support a central agenda. This type of arrangement provides the means for government to exercise power over other social actors at local or regional levels. This usually involves the delegation of government functions from the centre to the periphery in the process of allocating gains from water security. Egalitarian regimes are benefit-sharing arrangements based on the principle that all people are equal and deserve equal rights and opportunities. Such arrangements are usually framed as a matter of water justice and highlight the desirability of equitably sharing the benefits arising from the use of water resources. Distributive arrangements are benefit-sharing arrangements involving voluntary exchanges established to support what Ostrom refers to as *quid pro quo* relationships. These relationships are essentially reciprocal, in that a party offers a favour or advantage in return for something. As such, the voluntary arrangements are meant to provide positive incentives to facilitate and support water security.

A note on conflict resolution

It is important to note that conflict is inherent in most benefit-sharing regimes. Conflict situations may turn out to be unconstructive, polarized and even intractable. Usually, the challenge facing participants is not the elimination of conflict but how to functionally address it when it arises (Arnold et al. 2012). Although most water security initiatives in Africa are not frequently confronted by large-scale violence, they often involve lesser social tensions over water-allocation decisions. Government officials may be party to such conflict situations or called on to serve as mediators. Whatever the case, however, they need specific approaches and skills to tackle conflict situations in a constructive manner. Much of the benefit-sharing literature shows the diversity of participants' interests and power issues underlying water security initiatives. Powerful participants can choose not to participate in benefit-sharing agreements if the conflict situation is framed in a way that does not correspond with their interests or if they believe it is in their best interest to act individually (Arnold et al. 2012). Various conflict-resolution approaches exist (including cooperative problem-solving, mediation and arbitration), but negotiation as an approach is all the more relevant in the context of water security (Kaner, 2006). Negotiation involves voluntary efforts aimed at resolving conflicts that arise from competing needs and interests. Parties involved in specific negotiations seek agreement rather than resorting to force (Kaner, 2006). Interest-based negotiation procedures seek to reconcile the needs, desires and concerns of the parties involved, while power-based procedures determine who is more powerful, implying that the stronger party gets to determine the outcome (Arnold et al. 2012).

Study area and methods

Lower Zambezi National Park

In 1983, a section of the Zambezi River basin was declared Lower Zambezi National Park, primarily for the protection, management and sustainable use of river-based resources. The park lies on the north bank of the river in south-eastern Zambia (Figure 1). It is mainly used for game viewing, boating, canoeing safaris, walking safaris, photographic safaris and angling. Prior to the establishment of the park, the area operated as a private game reserve controlled directly by the office of Zambia's first and longest-serving president (Ministry of Tourism [MoT], 1992). This high-level, privileged control protected the area from the negative effects of mass tourism. Between 1975 and 1990, access to the area became difficult due to the political upheavals in the adjacent country, Northern Rhodesia (now Zimbabwe), where warring factions were involved in military

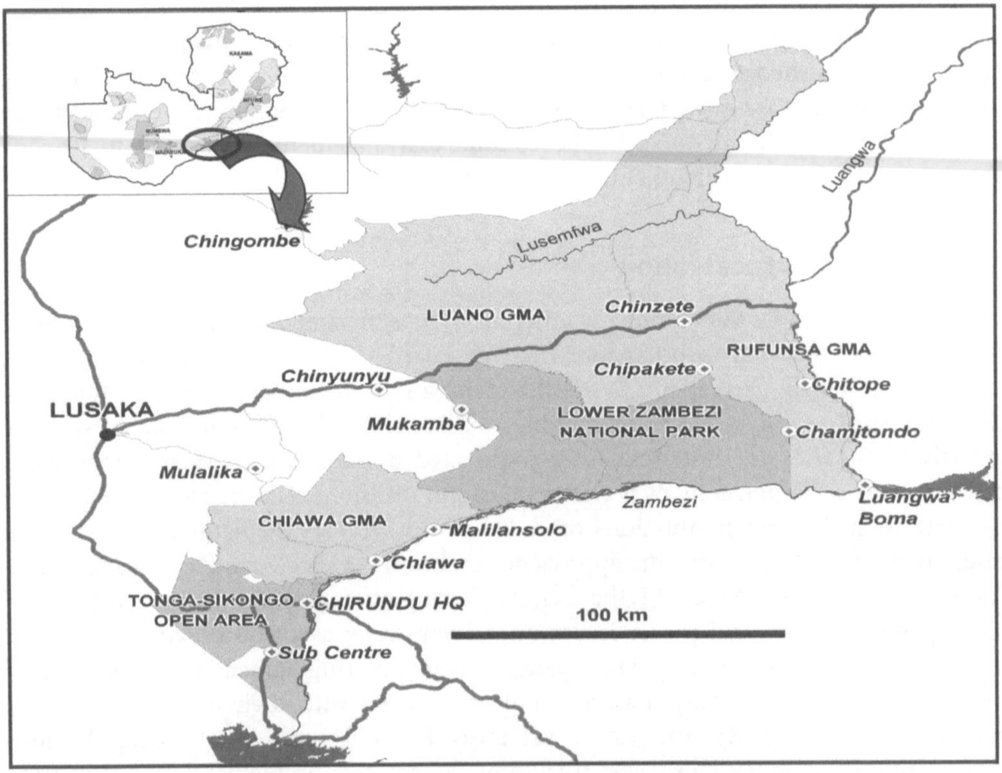

Figure 1. Map showing Zambia and the position of Lower Zambezi National Park.

conflict (Zambia Wildlife Authority [ZAWA], 2001). Few people would venture into the park for tourism-related activities for fear of military devices such as land mines and unexploded ordnance. Due in large part to this combination of factors, the area remained one of the few pristine wilderness areas left in Africa.

The restoration of tourism activities in the park began later, in 1990 (ZAWA, 2001). Interest in the area as a tourism destination was ignited after it was established that danger from land mines and unexploded ordnance left behind during the political upheavals was minimal. Tourism is considered the most important non-consumptive use of the Zambezi River in the park (ZAWA, 2001). However, most of the tourism-related activities in the park are restricted to a relatively small zone of the river frontage commonly referred to as the tourism zone, which is about 45 km^2 in size and extends for about 65 km along the river (Figure 2). This is in large part because the rugged escarpment lies close to the river. Despite the tourism zone being the smallest zone in the park, it still has to meet the ever-expanding demands for access to the river frontage from tourism operators and their visitors. The tourism operators mainly use camping sites along the river to host their clients.

Boating and canoeing are the major tourism activities in the tourism zone. The viewing of game and scenery from boats and canoes is the most popular activity, perhaps more popular than in any other park in Zambia (ZAWA, 2001). The Zambian side of the river has a number of narrow channels, which make for exciting game and scenery viewing. Fishing from boats is another popular activity, although all

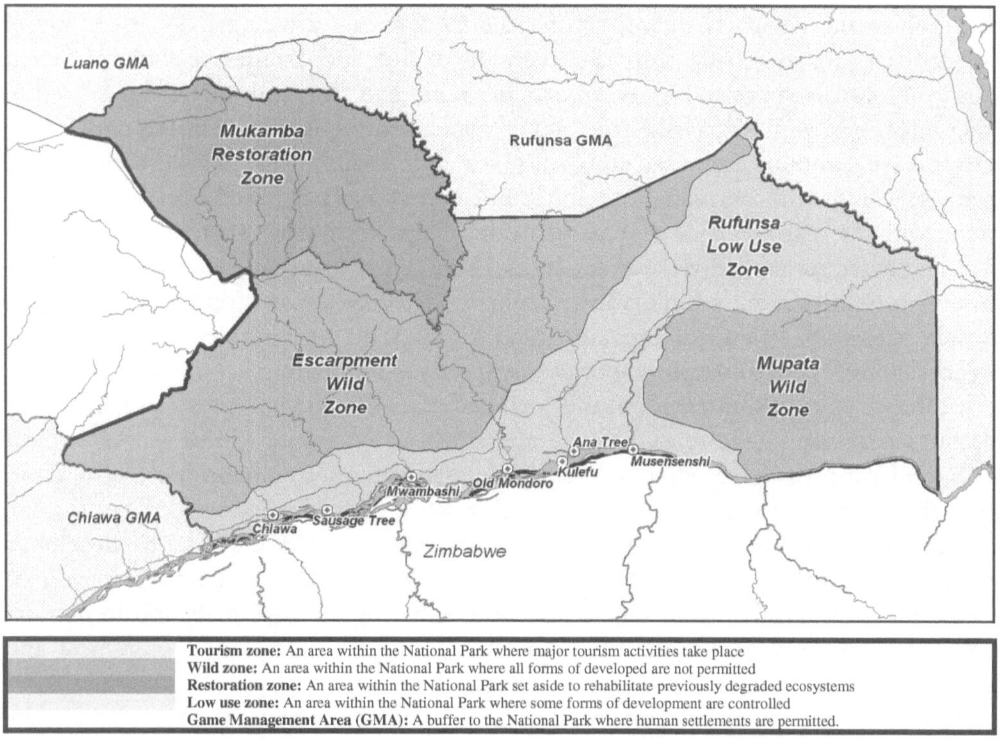

Figure 2. Management zones of Lower Zambezi National Park.

operators are required to adhere to a catch-and-release policy (ZAWA, 2001). The main species caught are bream and tiger. While on the Zimbabwean side of the river (Mana Pools National Park, a World Heritage Site) authorities have for many years banned the use of motor boats for tourism purposes, this type of regulation has not been widely accepted by authorities and operators on the Zambian side (ZAWA, 2001).

While clearly a major attraction in the tourism zone, boating is a major source of conflict between the operators with a permanent presence in the park and those based outside, in the adjacent Chiawa Game Management Area (GMA). The growing number of tourism facilities on the boundary of the park in Chiawa GMA has resulted in increased demand from outside operators for access to the tourism zone, which supports most of the available boating services in the park. The relatively smaller tourism zone is preferred for its spectacular scenery; Chiawa GMA is not as spectacular.

Methods

Qualitative field research methods, including interviews and documentary analysis (Strauss & Corbin, 1998), were used to explore the potential of benefit sharing in improving water security. A case-study approach, as described by Noor (2008), was used to develop a detailed account of the boating activities in the park. The delineation and scope of the case study were limited to boating activities, benefit-sharing events and mechanisms, and water security performance. The qualitative approach involved transcribing the interviews to gain a comprehensive understanding of the phenomena being studied. The study developed a data-coding method linking data codes to form themes, which led to the development of theory. The study focused on a single case study of boating to establish relationships between the codes and themes related to benefit sharing. Protocols suggested by Bradburn, Sudman, and Wansink (2004) were used to guide interviews with knowledgeable people about boating activities in the park.

Purposive sampling was used to identify key informants from two domains: inside operators (those with permanent facilities in the national park) and outside operators (those with facilities outside the park). With help from park staff, operators were identified based on their diverse experiences, management responsibilities and property ownership. Semi-structured interviews were conducted based on a purposive sample of 6 inside operators, 10 outside operators, and 4 park staff. An interview guide was used to generate historical and contemporary perspectives on boating activities in the park. Each interview was about an hour and included a visit to the operator's property. Key questions addressed were related to the operator's access options, preferences, experiences, and concerns over boating activities in the park. Respondents were also asked about their perceptions of opportunities and constraints for accessing the river system, particularly in the tourism zone. Interview transcripts were analyzed to develop a system for categorizing benefit-sharing regimes. The data from interviews were used to describe stages in the evolution of the benefit-sharing arrangements and to identify conflict-resolution interventions that were related to the themes of hierarchical and egalitarian benefit-sharing regimes.

To supplement the informant interviews, documentary material generated by the operators and other legal bodies associated with boating activities in Lower Zambezi National Park was analyzed. One major advantage of this approach was that the documents were generated contemporaneously with the events they referred to. As such, the information gathered was less likely to be subject to memory decay or memory distortion. A survey of documentary information relevant to boating activities had to be conducted both on site in the national park and at the headquarters of the Department of National Parks in Lusaka, the capital city of Zambia. The key sources of the information collected included the department's documentation, management decisions of the local park authorities, reports on park visits, and field reports on the official operations involving the operators. This information was captured from documentary platforms such as visitor data sets, reports, records of proceedings, minutes, briefs, and office records. These platforms were a useful source of information on the issues and events regarding the boating activities.

Evolution of benefit-sharing arrangements

In this section, I examine how the sharing of access to the benefits associated with boating activities in the tourism zone of the park evolved over the years. I specifically focus on two crucial periods that have defined benefit-sharing arrangements in the park: 1991–2000, which I have termed the hierarchical phase; and 2001–2010, which I refer to as the egalitarian phase. Later on I look at the implications and prospects of a distributive phase, which is currently being advocated by the inside operators. The aim is to demonstrate how these regimes have transformed over time from hierarchical to egalitarian benefit-sharing arrangements through the application of conflict-resolution mechanisms.

Hierarchical phase, 1991–2000

The hierarchical phase was largely dominated by a benefit-sharing regime designed to deal with the regulation of relationships among governmental actors and private tourism operators. This phase witnessed the resurgence of tourism in the park, largely driven by government's central agenda to diversify the country's economy, which was previously a mono-economy based on copper production (MoT, 1992). While the government was keen to develop the tourism sector, it did not have the financial capacity to put in the infrastructure required. It had to depend on the private sector for this. The opening up of tourism led to the bourgeoning of accommodation facilities established by private operators, both in the park and in the adjacent Chiawa GMA, with potentially dire consequences for local water security.

The total bed capacity in the park grew from virtually nothing in 1990 to 68 in 1992 (Table 2), distributed among 4 semi-permanent camps (48), two fly camps (12), and one camping site (8). All the accommodation facilities were run by licensed private operators who obtained accommodation leases from the Department of National Parks

Table 2. Bed capacity in Lower Zambezi National Park and the adjacent Chiawa Game Management Area, 1992–2010.

Area	1992	2000	2010
Lower Zambezi National Park	68	52	90
Chiawa Game Management Area	56	180	280

as the local river basin agency. Outside the park in Chiawa GMA, the number of operators had grown to five by 1992. Of these five, only two offered accommodation facilities in the GMA at permanent camp sites; the rest offered fly-camp facilities. The bed capacity of the GMA (56) then was smaller than the park's (68). All the revenue generated through the accommodation leases as well as boating permit fees went to the coffers of the Department of National Parks. Operators were permitted access to the boating services in the park on annual permits. As revealed by a respondent, all the accommodation facilities in the park and in Chiawa GMA were by and large established to facilitate access to the limited boating services in the tourism zone of the park (Tembo, 2016). Boating services in the GMA were not popular as the area did not have enough wildlife and fish stocks for a quality tourism experience (MoT, 1992).

Previously, outside operators were not permitted to access the boating services in the tourism zone of the park. But the government later realized that some planning was necessary to stop uncontrolled development and to ensure that access to the boating services was reasonably shared among all operators. In 1992, a benefit-sharing agreement was reached as a result of what can be conceived as power-based negotiations by the government agency in an effort to bring together outside and inside operators. The power-based negotiations were effected through face-to-face meetings facilitated by local government officials (Tembo, 2016). The resultant agreement allowed access to the boating services in the park by the outside operators. The stated goal of this benefit-sharing agreement was to provide access to areas 'where present and future generations of Zambians and other nationals can experience the inspirations of wild country' (MoT, 1992, p. 18). The agreement specifically indicated that this goal was to be consistent with the intent to limit development to specific zones, while at the same time ensuring that associated benefits were maximized for the benefit of the state, park and investors (inside operators). To achieve this broad goal, the agreement established some guidelines for the placement of operators and regulation of boating activities to ensure that the water security of the park was not degraded. This was to be achieved through controlling boating activities, to minimize river degradation and to ensure that the quality of the wilderness experience was paramount. The boating guidelines limited the number of boats that could be used by inside operators in the tourism zone to two per day. Outside operators were also allowed only two boats per day.

Although this agreement was viewed as a major milestone insofar as the underpinning power-based negotiations were concerned, its implementation over the years was problematic, with dire consequences for water security. Studies on the effects of boating on the Zambezi River showed that riverbank erosion had increased significantly, affecting water quality (ZAWA, 2001). In particular, boats with higher-horsepower motors affected water clarity by increasing the amount of sediment particles in the river. On the other hand, no sooner had the agreement been put in place than the government reversed the decision to give outside operators access to the boating services in the tourism zone. According to a respondent, the change in decision was a result of the influence of inside operators, who opposed the benefit-sharing agreement (Tembo, 2016). The inside operators argued that since they were solely responsible for maintaining the access roads in the park as well as contributing to the conservation of the river system, it was unfair to give outside operators, who never made similar contributions, access to the boating services in the park. The inside operators further argued that the benefit-sharing agreement led to overcrowding in the small tourism zone, which in their opinion interfered with the quality of the wilderness experience. Despite the merits of these arguments by the inside operators, it is important to note here that from the onset there was serious opposition from the inside operators to the increasing numbers of accommodation facilities outside the park, in the GMA (Tembo, 2016).

Subsequent efforts to implement boating guidelines were futile, as inside operators refused to recognize the right of outside operators to access the boating services inside the park. According to a government official who was interviewed, the outside operators were bent on flouting whatever guidelines were in place to regulate the boating activities (Tembo, 2016). Interviews conducted with the outside operators revealed that they were of the strong opinion that national parks were public property, and so access to benefits in these areas should be opened up to the greater public, not just a few selected individuals. Outside operators criticized the boating guidelines as not forming part of the 'agreement' in the true sense of the term as it did not allow for broader stakeholder consultations to capture the views of the outside operators. According to representatives of outsider operators, the consultative process focused on the needs of inside operators and selection of sites, efforts that were constrained by the situation on the ground as sites had already been identified by the existing inside operators. Prior to the agreement, inside operators had already chosen their favourable sites.

Failure to effectively implement the benefit-sharing agreement led to the further proliferation of tourism facilities on the boundary of the national park. The government's efforts to regulate tourism activities were further impeded by uncontrolled land allocation in Chiawa GMA, which was under customary tenure. Traditional leadership was responsible for land allocation in the GMA. While the existing laws indicated that the government had to be consulted for any land allocation in the GMA, evidence indicates that in many instances the traditional leaders never consulted the government. It appears that government officials at that time did not appreciate that many of the operators of the mushrooming facilities in the GMA wanted access to the park for boating activities. The inside operators were more concerned about overcrowding in the tourism zone. Unfortunately, the growing demands from outside operators and the concerns of overcrowding among inside operators were not adequately captured in the benefit-sharing agreement. As a consequence, the bed capacity of the GMA grew from 56 in 1992 to 180 in 2000, transforming the complexity of the benefit-sharing arrangements in the park. In contrast, the bed capacity of the park fell from 68 in 1992 to 52 in 2000 (Table 2). This was the result of the closure of one facility in the park. Generally, this scenario created conditions for renegotiating the benefit-sharing agreement and paved the way for the next phase of benefit-sharing arrangements in Lower Zambezi National Park.

Egalitarian phase, 2001–2010

The egalitarian phase was mainly dominated by a benefit-sharing regime designed to address past social injustices related to equitable access to the boating services in Lower Zambezi National Park. In particular, this phase saw the Zambian government put in place measures aimed at inculcating the principle that all resource users were equal and deserved equal rights and opportunities related to the use of boating services (Tembo, 2016). The launch of the phase was marked by the ratification of a new benefit-sharing agreement in 2001. The agreement was established through an interests-based negotiation tool commonly referred to as the strategic planning process (SPP) (ZAWA, 2001). The SPP was adopted in 1997 by the Department of National Parks premised on its potential to conflate the dynamics of water security requirements with interests-based negotiations that sought to reconcile the needs and concerns of the parties involved: government, outside operators and inside operators. Using the SPP, the Department of National Parks was able to establish the 2001 benefit-sharing agreement. According to Tembo (2016), the SPP was particularly important in proving the intrinsic strength of interests-based negotiations, which were used to encourage and offer parties opportunities to voice their concerns and interests regarding the management of Lower Zambezi National Park in general and the Zambezi River in particular. Opportunities for participation included three planning workshops, several technical meetings, informal work sessions, and public reviews of key draft documents.

The 2001 benefit-sharing agreement was facilitated by a qualified and appropriate multidisciplinary team of professional planners, who included specialists in fields such as river basin management, protected area planning, hydrology, extension work, sociology, and ecology. The team was ideally constituted in a manner that reflected an interdisciplinary approach. Managers from the park and the surrounding GMAs also had an important responsibility of actively participating in all the stages of the SPP. The professional planners together with the managers engaged in technical tasks such as literature review, field reconnaissance surveys, realigning management concerns, and objectives and action plans, as well as preparing the draft agreement. All these tasks epitomized the nature and substance of an adaptive interests-based negotiation approach to benefit sharing.

Of particular interest to this article, the 2001 benefit-sharing agreement attempted to address the demands of the outside operators regarding access to the boating services in the park. Following extensive interests-based negotiations, which involved both inside and outside operators, the parties were able to craft a benefit-sharing agreement in which four boats from outside were permitted to access the tourism zone on four days of the week. This agreement further reflected major progress in the interests-based negotiation efforts aimed at resolving the persistent contentious benefit-sharing arrangements in the park. While at the time both the inside and outside operators hailed the design of the agreement as a major achievement, its implementation had yet to be tested in reality.

No sooner was the 2001 benefit-sharing agreement ratified than the inside operators started complaining that the new boating guidelines were again being flouted by the outside operators. Several complaints were lodged by the inside operators suggesting that the boating guidelines were not working according to the benefit-sharing agreement (Tembo, 2016). The implementation of the agreement was again steeped in major controversy, with the two parties holding opposing views. The serious opposition from the inside operators that was experienced prior to 2001 resurfaced. This opposition was even extended to other tourism activities in the park. For example, the controversy culminated in 2004 in a court injunction which was brought by the inside operators, who were trying to stop the conversion of a non-operational public campsite into a lodge. The injunction prohibited any form of camping in the park and was invoked to bring the implementation of the agreement to a halt in late 2005. The injunction brought all further development in the park to a virtual standstill.

As indicated by respondents, conflict-resolution efforts were made by the Department of National Parks to try to convince the inside operators to drop the injunction. Several meetings were held involving both the inside and outside operators to resolve the impasse. Minutes from some of these meetings indicate that the inside operators were not happy with the enforcement mechanisms that were in place to regulate boating activities in the park. They noted that some outside operators were in fact taking advantage of the laxity of the enforcement mechanisms. They noted that a number of outside operators had gone so far as to use private boats to enter the park, to circumvent the new requirements. In response, officials from the Department of National Parks indicated that the 2001 benefit-sharing agreement was consistent with the limits of acceptable use based on credible baseline studies. They observed that these baseline studies took into account the impacts of tourism activities on the river system and the environment. However, the impasse continued into 2016, and as a result the much-needed benefit-sharing agreement could not be successfully implemented.

While this conflict was raging, uncoordinated development in the adjacent Chiawa GMA continued unabated. The bed capacity of the GMA grew from 180 in 2000 to 280 in 2010 (Table 2). In the park, the bed capacity also increased, from 52 in 2000 to 90 in 2010, probably as a competitive response to the developments in the GMA. To be sure, the management of these ever-changing, expanding and competing demands for boating services will probably continue to be among the greatest challenges facing the Department of National Parks as the local river basin agency in Lower Zambezi National Park.

Towards a distributive phase, 2011–?

At the beginning of 2011, the inside operators began advocating a distributive benefit-sharing arrangement which would potentially exclude outside operators from benefiting from the boating services in the park. The inside operators were looking at the prospects of designing and implementing a voluntary benefit-sharing arrangement that would provide positive incentives for them to sustainably manage and produce boating services. The inside operators were aiming to transform the tourism zone into an exclusive block would will cater to their clients only. Under the block-concession arrangement, the inside operators would assume a greater role in the planning, financing, design, construction, operation and maintenance of the infrastructure and tourist facilities in the exclusive block. In turn, the block would be completely closed to access by the other operators and the public, with the exception of the clients brought in by the inside operators. The inside operators hoped to establish a management trust that would administer the affairs of the exclusive block.

The idea behind the distributive benefit-sharing arrangements is to create river basin management incentives for the inside operators for actions that change resource-use patterns. This would entail offering benefits in excess of the costs incurred in the maintenance of the tourism zone as well as the related boating activities. It is envisaged that such an arrangement will assist in building sustained legitimacy for benefit-sharing initiatives in the park. Proponents of this benefit-sharing regime argued that if too many people benefit from something they have not actively contributed to or have no legitimate claims to, the associated incentives may be diluted. This would lead to weaker incentives and lower overall benefits to share. Thus, the 2011 distributive benefit-sharing arrangements sought to garner legitimacy by way of focusing on those people that were directly affected by the benefits.

Conclusion

This article sought to expose the role of benefit sharing in improving water security. In particular, it aimed to highlight the potential and limits of benefit sharing in buttressing local water security initiatives. The article has illustrated how the non-consumptive use of a section of the Zambezi River has shaped local water security over the years. It has critically examined how the sharing of benefits in Lower Zambezi National Park has evolved in the past 30 years, with implications for local water security. The implications of increasing water insecurity were epitomized in studies on the effects of boating on the Zambezi River that showed that riverbank erosion had increased significantly, affecting water quality. From the foregoing narrative, it is tempting to suggest that local water security has not materially improved. But this study has made an effort to go beyond common issues of maximizing water availability and minimizing water risks to also focus on how conflict shapes the transformation of benefit-sharing arrangements and water-allocation processes. The findings of this study clearly indicate that the benefit-sharing arrangements in the park have transformed through negotiations from a hierarchical to an egalitarian regime.

The inherent conflict between the inside and outside operators was a major source of transformation. The growing number of tourism facilities on the park boundary led to increased demand for use of benefits associated with boating activities. Initially, outsider operators were not permitted to benefit from the boating activities. However, as a result of ongoing conflict-resolution efforts, a benefit-sharing agreement was reached in 2001. The analysis of the evolution of this benefit-sharing agreement driven by conflict raises important questions about the relationship between benefit sharing and water security. A key question is whether current-benefit sharing models are well suited to deal with the conflict inherent in local water security initiatives.

The efforts to apply conflict-resolution mechanisms, primarily through power-based negotiations in the hierarchical phase, were largely unsuccessful due to poor design and weak implementation of the benefit-sharing agreement. Clearly, sufficient mechanisms were not put in place to deal explicitly with persistent conflict. For example, there was no comprehensible monitoring plan to ensure the tracking of major decisions and their implementation. And it is apparent that the relationship between outside and inside operators was not effectively managed. Importantly, there was limited stakeholder engagement in this phase. As a result, the ensuing conflict-resolution efforts were not adequately directed. This led to a situation where the outside operators viewed the tourism zone as private property, whereas the outside operators saw it as public.

Coming to the egalitarian phase, it is clear that the effort to apply interests-based negotiations produced mixed results: an improved design but weak implementation. On the other hand, a robust monitoring system was put in place to deal with conflicts. The management of the relationship between outside and inside operators was also improved, as evidenced by a strong focus on interests and extensive stakeholder engagement. But these conflict-resolution efforts were still unsuccessful, as evidenced by the various meetings that were held. Ironically, the inside operators continued to view the tourism zone as private property, and the outside operators maintained their perception of it as public.

This case study thus illustrates that benefit sharing in river basins occurs in complex, uncertain and contested contexts. Decisions regarding benefit sharing involve complex value judgements, and decisions on sharing options usually lead to controversy and conflict. As illustrated in the case study, decision-making processes around benefit sharing often suffer from crises of illegitimacy. Thus, the explicit and deliberate use of conflict-resolution frameworks is vital. The case study clearly demonstrates that benefit sharing in river basins is founded on complex long-term relationships that involve multiple interest groups with divergent expectations and experiences. These relationships are usually characterized by behavioural processes in which interest groups influence each other's behaviours over a period of time to advance individual and shared goals. The nature of such relationships is critical in determining the success or failure of benefit-sharing schemes insofar as efforts to improve water security are concerned. It is essential that water security approaches take sufficient account of the nature and substance of the long-term relationships that underlie particular benefit-sharing schemes.

Acknowledgments

This work was supported by the International Water Security Network which is funded by Lloyd's Register Foundation, a charitable foundation helping to protect life and property by supporting engineering-related education, public engagement and the application of research.

Disclosure statement

No potential conflict of interest was reported by the author.

References

Angelakis, A. N., Mays, L. W., Koutsoyiannis, D., & Mamassis, N. (2012). *Evolution of water supply through the Millennia.* London: IWA Publishing.

Arnold, J. S., Koro-Ljungberg, M., & Bartels, W. (2012). Power and conflict in adaptive management: Analyzing the discourse of riparian management on public lands. *Ecology and Society, 17*(1), 19.

Boyle, M., Kay, J., & Pond, B. (2001). Monitoring in support of policy: An adaptive ecosystem approach. In T. Munn (ed.), *Encylopedia of global environmental change* (Vol. 4, pp. 116–137). New York: John Wiley and Son.

Bradburn, N. M., Sudman, S., & Wansink, B. (2004). *Asking questions: The definitive guide to questionnaire design—For market research, political polls, and social and health questionnaires.* San Francisco, CA: Jossey-Bass.

Chomba, M. J., Hill, T., Nkhata, B. A., & Förster, J. J. (2017). Paradigms for water allocation in river basins: A society-science practice perspective from Southern Africa. *Water Policy, 19*, 637–649.

Convention on Biological Diversity. 1992. *Convention on biological diversity.* Retrieved 3 October, 2018, from https://www.cbd.int/doc/legal/cbd-en.pdf.

Fischhendler, I. (2013). The securitization of water discourse: Theoretical foundations, research gaps and objectives of the special issue. *International Environmental Agreements, 15*(3), 245–255.

Global Water Partnership. (2000). *Towards water security: A framework for action.* Stockholm: Author.

Grey, D, & Sadoff, C.W. (2007). Sink or swim? water security for growth and development. *Water Policy, 9*, 545–571.

Hayden, C. (2003). From market to market: Bioprospecting's idioms of inclusion. *American Ethnologist, 30*(3), 1–13.

Hoekstra, A. Y., Buurman, J., & van Ginkel, K. C. H. (2018). Urban water security: A review. *Environmental Research Letters, 13*, 053002.

Kaner, S. (2006). *Facilitator's guide to participatory decision-making.* San Francisco, CA: Jossey-Bass.

Mekonnen, D. Z. (2010). The Nile Basin cooperative framework agreement negotiations and the adoption of a 'water security' paradigm: Flight into obscurity or a logical cul-de-sac? *European Journal of International Law, 21*, 421–440.

Ministry of Tourism. (1992). *General Management Plan for Lower Zambezi National Park.* Lusaka, Zambia: Government of the Republic of Zambia, Ministry of Tourism.

Nkhata, B. A., Breen, C. M., & Mosimane, A. (2012a). Engaging common property theory: Implications for benefit sharing research in developing countries. *International Journal of the Commons, 6*(1), 52–69.

Nkhata, B. A., Mosimane, A., Downsborough, L., Breen, C., & Roux, D. J. (2012b). A typology of benefit sharing arrangements for the governance of social-ecological systems in developing countries. *Ecology and Society, 17*(1), 17.

Noor, K. B. M. (2008). Case study: A strategic research methodology. *American Journal of Applied Sciences, 5*(11), 1602–1604.

Ostrom, E. (2005). *Understanding institutional diversity.* Princeton, NJ: Princeton University Press.

Pahl-Wostl, C., Lebel, L., Knieper, C., & Nikitina, E. (2012). From applying panaceas to mastering complexity: Toward adaptive water governance in river basins. *Environmental Science & Policy, 23*, 24–34.

Petersen-Perlman, J. D., Veilleux, J. C., Zentner, M., & Wolf, A. T. (2012). Case studies on water security: Analysis of system complexity and the role of institutions. *Journal of Contemporary Water Research & Education, 149*, 4–12.

Strauss, A, & Corbin, J. (1998). *Basics of qualitative research: Techniques and procedures for developing grounded theory.* Thousand Oaks, CA: Sage Publications.

Tembo, S. (2016). *Personal communication.* Lusaka, Zambia: Mr. Tembo was the government official responsible for tourism in Lower Zambezi National Park from 2006 to 2016.

Turton, A. (2008). A South African perspective on a possible benefit sharing approach for transboundary waters in the SADC Region. *Water Alternatives, 1*(2), 180–200.

United Nations University/Institute for Water, Environment & Health. 2013. *Water Security and the Global Water Agenda: A UN-Water Analytical Brief.* 175 Longwood Rd. South, suite 204, Hamilton, Ontario L8P 0A1 Canada.

Vatn, A. (2010). An institutional analysis of payments for environmental services. *Ecological Economics, 69*, 1245–1252.

Zambia Wildlife Authority. (2001). *General management plan of Lower Zambezi National Park, 2001* (Vol. 1). Chilanga, Zambia: Author, Private Bag.

Zeitoun, M. (2011). The global web of national water security. *Global Policy, 2*, 286–296.

OPEN ACCESS

Why doesn't every family practice rainwater harvesting? Factors that affect the decision to adopt rainwater harvesting as a household water security strategy in central Uganda

Chad Staddon, Josh Rogers, Calum Warriner, Sarah Ward and Wayne Powell

ABSTRACT

This article investigates the reasons householders do, and don't, adopt domestic rainwater harvesting (DRWH). Using a mixed-methods research approach, we collected data in three districts in central Uganda. Factors that emerged as important with respect to uptake of DWRH to address water shortage, especially at the household scale, include the work of intermediary organizations, finance mechanisms, life course dynamics and land tenure.

Introduction

Notwithstanding two 'Water and Sanitation Decades' and innumerable global conferences, declarations and programmes, more than 800 million people in developing nations still lack even the most basic levels of personal or household-scale water security (AMCOW, 2008; Siwar & Ahmed, 2014; WHO, 2017; WHO/UNICEF, 2015). Although water is essential for maintaining life and for all social and economic endeavours, several authoritative global reports (e.g., World Water Assessment Programme, 2018; UN-Water, 2015) have suggested that urbanization, population growth, industrialization and increasing per capita consumption of all goods including water mean that by 2030 there could be a 40% global water deficit. In some developing regions – especially sub-Saharan Africa and Latin America – water security will actually deteriorate rather than improve, despite decades of investment in water and sanitation infrastructure (WWAP, 2018). Improving water security at any scale (household, urban, national, etc.) presents a massive challenge in a rapidly growing and globalizing world where local needs might not just outstrip local water resources, but where those resources can also 'travel' in the form of physical or embodied exports to other regions (e.g., virtual water export via the global food and horticultural trades, see Allan, 2011; Staddon, 2010). With amplified pressures on water governance through increasing transboundary tensions (explored further by Albrecht et al. in this collection), the deregulatory and laissez-faire impulses of the neoliberal Washington Consensus, climate change uncertainties and anthropogenic pressures, the already severe threat to water security in the Global South is continuing to grow (Bakker, 2012; Lovelle, 2016; Owen & Goldin, 2015).

This article focuses on challenges to improving water security in one developing nation, Uganda. Uganda's economic and infrastructure development has been dramatically affected by frequent periods of violent conflict and political repression since independence in 1962. In this context, it is perhaps not surprising that the first National Water Policy was not passed until 1999 (Nsubuga, Namutebi, & Nsubuga-Ssenfuma, 2014; Terry, McLaughlin, & Kazooba, 2015). Since then the country has made huge progress in restructuring and developing water supply and sanitation throughout the country, with access to at least 'improved water sources' at 95% for urban areas and 71% for rural areas by 2015 (WHO/UNICEF, 2015). Unfortunately, as noted above, 'improved water sources' does not necessarily mean that water is either proximate to, or sufficiently clean and plentiful for, all households all of the time, particularly in the poorest and most remote districts.[1] Moreover, there remain significant urban/rural, regional and other disparities. While estimates suggest that 71% of the urban population have some sort of access to potable water managed by the National Water and Sewerage Corporation, this is only about 14% of the total population. The other 86% of the Ugandan population depends on 'other improved' or 'unimproved' sources, which may include local surface water sources, hand-dug shallow wells or groundwater collection schemes (Figure 1; Nayebare, Wilson, Carpenter, Dziewulski, & Kannan, 2014; WHO/UNICEF, 2015).

While many analysts urge policy and practical action at national and international scales, there is much that can be done to improve water security at household and personal scales. Moreover, as Wade argues in this collection, water security needs to be understood as a *scale concept* so that potentially deleterious scale interactions can be mapped and, ideally, avoided. One strategy that has proven useful in addressing household-scale water insecurity is the installation of local storage systems that take in water from specially prepared collection/catchment areas (usually roofs, but sometimes also areas of ground – see Figure 2), and save it in vessels for later use (Abdulla & Al

Figure 1. 'Unimproved' (left) and 'improved' (right) water sources. Photos by the authors.

Figure 2. Schematic indicating basic configuration of domestic rainwater harvesting systems in Uganda. Source: the authors.

Shareef, 2009; Ahmed, Mustafa, & Khalid, 2011; Mati, 2007). Such domestic rainwater harvesting (DRWH) systems are relatively easy to build with local skills and resources and can be commissioned in households at risk of water insecurity relatively quickly, though they do bring with them challenges of appropriate sizing to meet demand under future climate uncertainty, asset maintenance and water quality management. As of June 2017, throughout Uganda there were approximately 21,000 rainwater harvesting tanks serving 126,000 people, around 1% of the rural population (Funk et al., 2005; Ministry of Water and Environment [MWE], 2017; UNFCC, 2007). Significant upscaling of appropriately sized DRWH systems will be required to meet future demands and safeguard communities facing the threat of water that is scarce and of inadequate quality. Building on ongoing research in Uganda (Terry et al., 2015), in this article we report on recent research into the reasons householders do, and don't, adopt DRWH. Using a mixed-methods research approach we collected data in three districts in central Uganda: Luweero, Wakiso and Mukono (Figure 3). The three study areas were chosen for relative ease of access to the capital and their contrasting experience with DRWH schemes. Factors that emerged as important with respect to uptake of DWRH to address water shortage, especially at the household scale, include the work of intermediary organizations, finance mechanisms, life course dynamics and land tenure. Better understanding of the factors influencing adoption of DRWH should assist in formulation of better policy by governments and development NGOs in Uganda and elsewhere in the developing world.

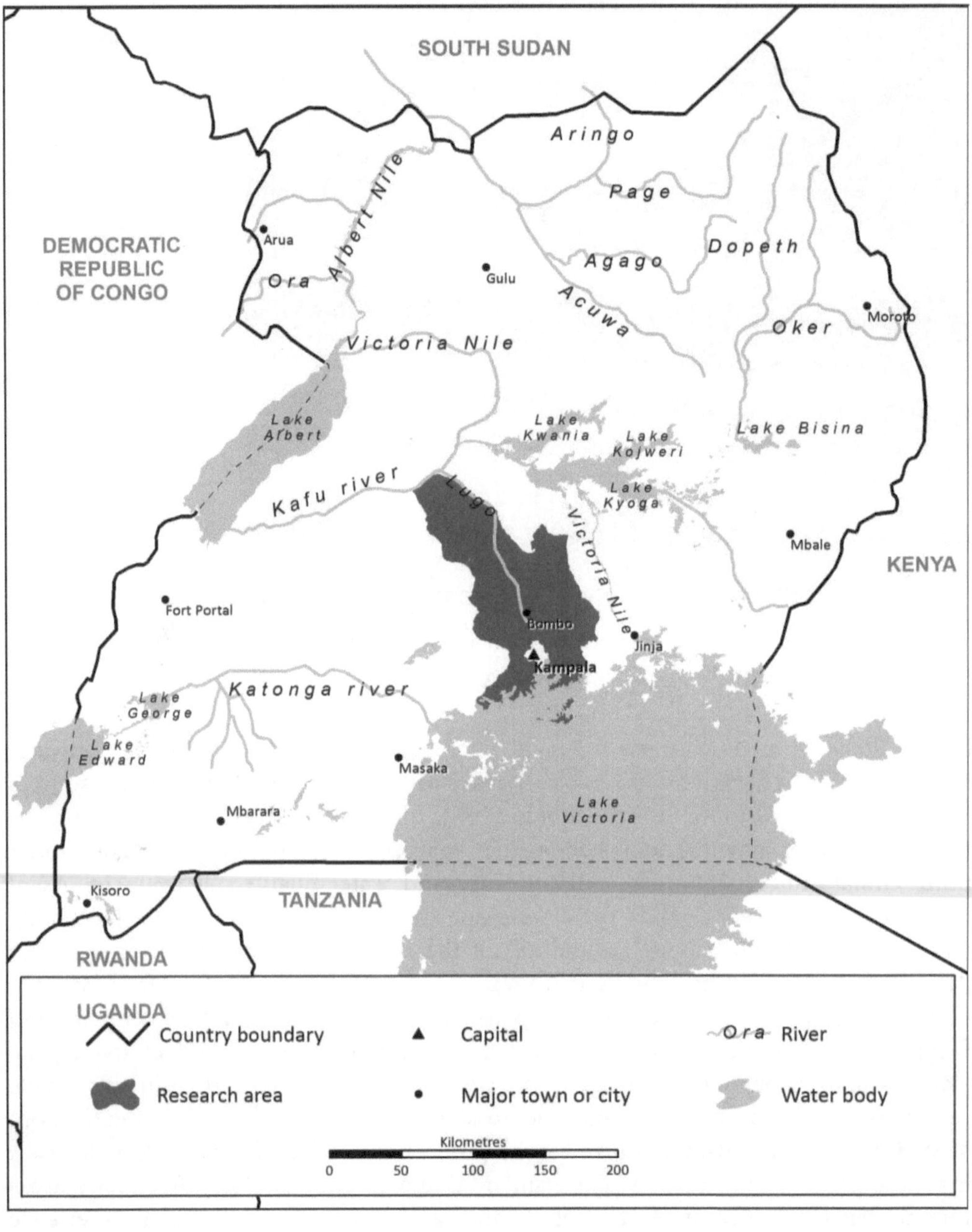

Figure 3. Map of Uganda indicating key study areas north of Kampala. Source: the authors.

Water security and DRWH

The materials and design of DRWH systems can be quite simple, though care needs to be taken in commissioning and operation to ensure a long life for the asset. For example, poor concrete compaction or incorrect sand–cement mixture in either foundations or tank walls (for ferro-concrete systems) can quickly compromise the entire system (Danert & Motts, 2009; Gould & Nissen-Petersen, 1999; Turner, 1999). DRWH is regarded in the development and water policy literature and by the Millennium Development Goals and Sustainable Development Goals as an 'improved water source' with great potential to provide additional water collection and storage for households in areas of unpredictable rainfall and beyond the reach of centralized reticulation systems (Kahinda, Taigbenu, & Boroto, 2007; MWE, 2014, 2017: Thayil-Blanchard & Mihelcic, 2015). This is important as it means that investments in expansion of DRWH count towards meeting the targets agreed by national governments for water and sanitation as part of the Millennium Development Goals (up to 2015) and the Sustainable Development Goals (through to 2030). In particular, Sustainable Development Goal 6 – clean water and sanitation – commits the governments to the challenging goal of universal drinking water access by 2030:

> Ensuring universal access to safe and affordable drinking water for all by 2030 requires that we invest in adequate infrastructure, provide sanitation facilities, and encourage hygiene at every level. Protecting and restoring water-related ecosystems such as forests, mountains, wetlands and rivers is essential if we are to mitigate water scarcity. More international cooperation is also needed to encourage water efficiency and support treatment technologies in developing countries. (UNDP, 2018)

Baguma and Loiskandl (2010) identify DRWH as a method that can mitigate water shortages, providing additional supply for personal, domestic and local gardening (food security) uses, and Ghimire and Johnston (2015) point out that there is a long tradition of using it. The Ugandan government's 'self-supply' policy supplements conventional centralized or semi-centralized water systems by encouraging households to access water through their efforts (Danert & Motts, 2009; MWE, 2010; Terry et al., 2015). Householders are encouraged to make incremental improvements to their supply using local and easy-to-replicate solutions, including particularly DRWH and, where conditions allow, shallow wells. Smet (2003), Pullan, Freeman, Gething, and Brooker (2014) and others suggest that drivers of the technology in the Global South include: a shift towards sustainable, community-based approaches to water supply; the failure of centralized water supply systems (the often chimeric dream of centralized piped supply); better availability of cheap tanks (especially ferro-concrete and HDPE plastic); and a decrease in the quality and quantity of ambient ground and surface water. Ugandan government policy in this area further notes the long history of self-supply reaching back into the colonial era as a foundation of this approach (MWE, 2010). Although a significant body of research exists with regard to the technical utility of rainwater harvesting to address household water insecurity, research into the reasons for adoption of DRWH in developing countries is more limited. Baguma and Loiskandl (2010, p. 356) assert that 'a knowledge gap exists on why some adopt or reject DRWH technologies and adoption trends continue to be inadequately understood at a household level'.

DRWH systems are a good example of a 'socio-technical' intervention (Linton, 2010; Staddon, 2010), depending as they do on the right mix of both physical and social conditions. Jepson et al. (2017) write that instead of seeing water as only a static physical object that should be 'captured' through clever civil engineering, household water security should be conceptualized as emerging from a 'hydro-social' process. In Figure 2 the basic components of a typical DRWH system are shown. The roof, usually corrugated iron (often this is a requirement of organizations promoting DRWH), is the collecting surface, whose collected water is directed via the guttering to the storage system. Thus, the typical system provides an above-ground storage facility for water for household use. Tanks are normally sized at between 4 and 8 kilolitres, with a view to being filled during wet seasons and therefore providing domestic water security during dry seasons. In this part of Uganda there are two wet seasons and two dry seasons each year. These systems are socio-technical in the sense that successful operation requires owners to undertake regular inspection and basic maintenance and indeed to understand that the water provided will change in quality (deteriorate) from the moment they are filled. This will necessitate attention to processing through on-site filtration and/or disinfection.

The relational approach outlined by Jepson et al. sees water security as a *process* that does not simply capture increments of physical water to satisfy demand (objectively understood in terms of litres per capita per day) but recognizes and understands the wider interactions that take place between people and water. Similarly, Linton (2010) suggests that access to water as a human right should be re-formulated as *the right to be involved in decisions that affect the way in which people and water interact.*[2] Applying this more relational perspective on household water insecurity derived from Jepson et al. (2017), Staddon and James (2014) and others, it is possible to analyze the cultural, social and political relations – what they refer to as the 'human capabilities' – *behind* the securing (or not) of the services we require from water (e.g., hydration, cleanliness, etc.). In other words, the hydro-social model of defining water security in relation to DRWH adoption trends takes into account households' interactions with other actors (private suppliers, government agencies, NGOs, the physical environment, etc.) that have led to securing water, or not securing it, for different purposes. For example, membership in intermediary community organizations or NGOs (such as those promoting DRWH and discussed in this article), access to appropriate technical knowledge and finance mechanisms, or access to a particular level of education may all be examples of wider relations that influence a household's access to water and that the hydro-social conceptualization takes into account (Baguma & Loiskandl, 2010; Jepson et al., 2017). In this sense, achieving water security is seen as a very dynamic process, bridging the gap between socio-political relations on the one hand and the acquisition of physical water on the other.

There are many different ways of defining and measuring water security in the academic literature (see Bakker, 2012; Scott et al., 2013; Staddon & James, 2014; Sun, Staddon, & Chen, 2016 for reviews). Webb and Iskandraani (1998, p. 4) offer the following definition: 'access by all individuals at all times to sufficient safe water for health and productive life'. Other definitions, such as that promoted by Scott et al. (2013), specifically incorporate the natural environment as a water user whose needs must be properly accounted. One simple way of measuring water security is by calculating a demand over supply ratio (e.g., Lanka Rainwater Harvesting Forum, 2001). Thus, where the demand for water outstrips the supply the ratio will be more than 1, and where the demand is smaller than the supply then the household's water security is less than 1. Ariyabandu (2001) suggests that the accessibility, reliability and timely availability of adequate safe water for basic human needs defines water security and that there is also a clear difference between 'perceptional quality' and 'absolute quality', meaning that water security may also be defined with reference to user attitude, perception and judgement. The 'capabilities' approach that we adopt from Jepson et al. (2017) and Wutich et al. (2017) helps us focus on key facilitating intersectionalities related, for example, to 'health and well-being', 'income and resource', and 'autonomy and self-determination'. There is considerable evidence that access to water services is positively correlated with reduction in global health burden (Forouzanfar et al., 2015). Moreover, it is important that we are sensitive to the complex interrelationships between water security policy and practice on the one hand, and socio-cultural factors that may influence water management and allocation (e.g., gifting and reciprocity) on the other.

Located approximately 75 km north of the capital (Figure 3), Luweero's economy is largely devoted to agriculture, both animal husbandry and crop production. Luwero is the smallest and most rural of the three study districts (Table 1). Its chief agricultural products are bananas, coffee, sweet potatoes and maize, though volumes of other products, such as honey, have been growing quickly in recent years (Uganda Bureau of Statistics, 2017a). Unfortunately, economic growth has also led to greater environmental degradation, particularly with respect to lost wetlands and forest cover, topsoil loss and water pollution. Housing is reasonably good, with 65% owning their own homes, 82% of which have corrugated iron roofs. Access to water resources is less salutary, with only 54% reporting access to safe water and 60% access to a pit latrine.

Demographic data for Wakiso and Mukono Districts, located immediately west and east of the capital, respectively, are also shown in Table 1. Given their peri-urban status (and therefore better access to the capital) most indices of socio-economic development are higher, including smaller households, higher levels of home ownership and better health statistics (Uganda Bureau of Statistics, 2017b). Consonant with this relative economic prosperity, and proximity to the capital, fewer residents than the national average (27%, compared to 68%) are dependent on subsistence farming. Of the three districts surveyed, Wakiso also has the best statistics for housing quality, availability of piped water supply (51.2%), improved sanitation (99.2%), and access to electricity (62%), but the lowest level of home ownership (44%). In a survey published by the *Daily Monitor* (Kafeero, 2018), Wakiso emerged as the country's richest district, with an average GDP per capita of US$ 3250, compared with US$ 57 in the poorest district, Kagada (in the Western District, near Lake Albert). Mukono has somewhat higher levels of home ownership than Wakiso, but lower levels of access to electricity and a higher proportion of households practising subsistence agriculture. While access to

Table 1. Data for the three study districts (Uganda Bureau of Statistics, 2017a, 2017b, 2017c).

District	Location	Total population	Rural	Growth	Area (km^2)	Population density (per km^2)	No. of households
Luwero	Central Uganda (North of Kampala)	458,158	79%	2.45%	2,557.5	17.7	106,235
Mukono	Central Uganda (East of Kampala)	599,817	73%	2.91%	10,211	58.7	145,575
Wakiso	Central Uganda (West of Kampala)	2,007,700	68%	6.10%	2,807.7	71.5	504,602

sanitation is similar to Wakiso, access to piped water supply is only 22% (Uganda Bureau of Statistics, 2017c).

Methodology

Using a multi-methods approach, this study evaluated adoption trends of DRWH in three rural districts in central Uganda, all within a short drive of the capital, Kampala (Figure 3). Completed over a three-month period in late 2015 and early 2016, the study compares capabilities variables related to household characteristics – such as education, age, access to finance, and membership in an intermediary NGO or community organization – to assess factors correlated with adoption of DRWH. Questionnaire surveys, focus groups and direct observations were all used to collect primary data as part of an integrated approach to data collection and triangulation (Cresswell, 2002; Danielson, Tuler, Santos, Webler, & Chess, 2012; Flick, 2014). The districts chosen for study are shown in Figure 3, and basic socio-economic data for these districts are given in Table 1. NGOs that acted as bridging or linking organizations between the research team and the local communities in the three districts included:

- the Busoga Trust, a community-based NGO working to provide sustainable access to safe water, improved sanitation, and hygiene and health education
- the National Association of Professional Environmentalists, an advocacy organization working closely with communities on economic and environmental issues
- the Katosi Women's Development Trust, an NGO working with women to enable them to improve their social, economic and political conditions.

Working with these organizations enabled the researchers to quickly identify useful study sites and interviewees and gave an easy entrée into these communities for the (white European) researchers. Researchers were also sensitive to the information needs of the partner organizations and sought to assist them wherever possible.

In the first research phase, recent secondary data (from government and other official sources) was collected to provide an up-to-date view of the landscape of DRWH in Uganda generally and the study districts in particular. A limited number of focus groups were also conducted as a way of introducing the researchers to the communities and gauging likely participation rates in the subsequent questionnaire survey phase. An initial, NGO-facing, focus group was organized by the Uganda Rainwater Harvesting Association during a two-day public dialogue in Gulu, northern Uganda, attended by 12 representatives of local government and NGOs. As Danielson et al. (2012) point out, focus groups are a kind of group interview that capitalizes on communication between research participants in order to generate data. This means that instead of the researcher asking each person to respond to a question in turn, people are encouraged to talk to one another, with the researcher noting areas of concord and discord. One key question posed to this group was: 'In your collective experience, what variables impact the decision of households to adopt or not adopt DRWH?' The responses given broadly fell into the five spheres of socio-cultural, socio-economic, financial, attitude, and perception, which were then used to structure the subsequent questionnaire survey.

In the second phase, the research deployed a semi-structured questionnaire survey that focussed on basic household characteristics, experiences of water insecurity and factors affecting adoption or non-adoption of DRWH. A key motivation behind the survey was the need to better understand the complex mix of enablements and constraints that face householders considering DRWH implementation. In all phases of the questionnaire process we were mindful of the dangers of various kinds of bias, particularly social acceptability and acquiescence bias (Gillham, 2008; Staddon & Genchev, 2013; Valentine, 1999). One question in each section of the questionnaire used a Likert scale to gauge the strength of respondents' opinions (Albaum, 1997). It was decided to present the questions verbally, with the use of a translator, due to low literacy rates and also to allow further discussion of open-ended questions. Upon arrival at a household there would be a brief introduction and explanation of the research to gain permission to interview the respondent and observe the respondent's homestead. The questionnaire was then conducted, after which respondents would give a tour of household water facilities. Approximately 25 minutes was spent at each household, although this time could vary significantly. In the third phase, interviews and focus groups with householders were used to add depth, detail and richness to the picture emerging from the earlier research phases. This helped contextualize quantitative understanding of the DRWH situation in Uganda by allowing the researchers to see for themselves the situations that respondents were describing, and also allowing respondents to draw attention to salient aspects of their domestic environments (Sin, 2003).

Examining the drivers of DRWH adoption trends involved surveying household characteristics, and for the purpose of this study it was assumed – guided by previous studies of this kind (Ahmed, Onwonga, Mburu, & Elhadi, 2013; Baguma & Loiskandl, 2010; Zingiro, Okello, & Guthiga, 2014) – that the self-declared head of household (usually the one home when researchers visited) was the most knowledgeable about DRWH issues. Initially, a 'random walk' method was chosen for sampling, starting with a 'seed value' household and then proceeding to every *x*th or *y*th household (Flynn, Tremblay, RFehm, & Wells, 2013). Since enumerators often encountered multiple heads of household together (e.g., neighbours visiting) we elected to move to a 'snowball' sampling method, acceptable since the statistical representativeness of the sample was not the primary desired outcome.[3] The intention was rather the capture of a sense of local lived experience of water insecurity and the creation of a platform for deeper explorations of household-scale strategies for addressing it, with particular emphasis on DRWH adoption. The achieved sample size of 66 was large enough to enable limited descriptive and analytical statistical examination, though (as noted above) not with the intention of necessarily speaking statistically about the larger population (Rogerson, 2010). Completed surveys were received from 12 villages in the three districts, with very few refusals from householders, and indeed a high level of enthusiasm, as householders are very interested in anything that may help improve the local water situation.

Taken together, the focus groups and household questionnaires tell a compelling story about why some families in some locations are more likely to adopt DRWH as part of their household water security strategy than other families in other locations.

Results and discussion

This section presents the results of the household survey, focus groups, and informal interviews and observations, focussing particularly on the stated reasons for adoption/

non-adoption of DRWH. Education, gender, age (of head of household), family size, membership in relevant intermediary organizations and access to credit were all considered. Overall, we found that gender is a better predictor of adoption than education, although neither presents a particularly strong association, whereas, age, family size, membership in a relevant organization, access to credit, land tenure, and attitude all have a significant positive relationship with adoption.

Importance of intermediary organizations

The strongest correlation was between membership in a community or intermediary organization and DRWH adoption, consistent with previous studies (Ahmed et al., 2013; Baguma & Loiskandl, 2010; Zingiro et al., 2014). Figure 4 shows the clear contrast between adopters and non-adopters, with 82% of adopters declaring membership in a community or intermediary organization, while only 7% of non-adopters are members. However, membership in a community or intermediary organization did not entail access to all other necessary enablers of adoption. In other words, membership seems to be a necessary but not sufficient condition for adoption. An NGO representative interviewed made it clear that they do not provide cash subsidies to households, instead opting to assist through supplying information, training and materials that may not be available locally, such as pipes, cement and guttering. This may facilitate adoption better than cash subsidies on their own, as there is evidence that cash subsidies are often used inappropriately (Baguma & Loiskandl, 2010). Following Jepson et al. (2017), it can be asserted that membership in a relevant community organization or NGO boosts household capabilities in the area of DRWH through facilitation of access to information, practical assistance, materials and even encouragement and moral support

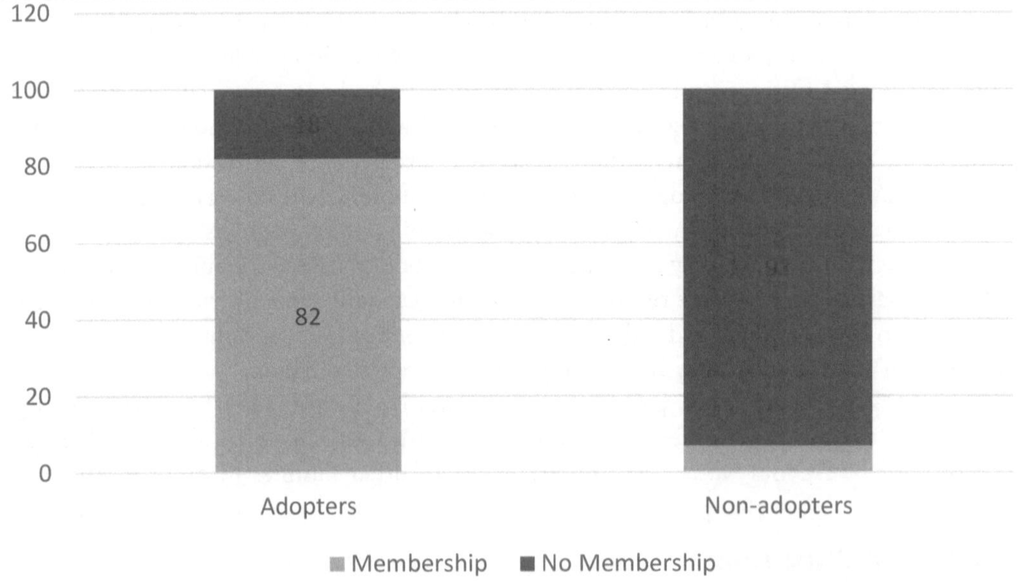

Figure 4. Percentage of DRWH adopters and non-adopters who are also members of an enabling organization/NGO. Source: the authors.

(e.g., for HIV widows). From the capabilities point of view, community and non-governmental organizations need to work closely with communities to exchange knowledge and improve uptake of DRWH, including through practical action.

Technical knowledge transfer of the benefits associated with DRWH may be key in this respect (Ahmed et al., 2013). During a public dialogue attended by the researchers in the northern city of Gulu, a representative from the Appropriate Technology Centre (an action research initiative established by the Ministry of Water and Environment, with headquarters in Mukono, just east of the capital) highlighted the need for technical improvements to the operation of DRWH tanks. These included closer attention to fill levels, which affect wetting/drying of internal surfaces, and better management of the limnologic stratification of the enclosed water, which is critical to the production of cleaner water. Ongoing research elsewhere in Uganda is focussing on the finer points of tank construction, including sand–cement ratios, sand types used, setting/curing time, etc., as variables that are strongly correlated with effective performance but which can be poorly understood by local users (Danert & Motts, 2009; Thayil-Blanchard & Mihelcic, 2015).

In a keynote address at the same event in Gulu, a representative of the Ugandan government suggested that DRWH presents a sustainable solution to rural water security issues if implemented by government on an institutional level, where standards can be appropriately policed. This perspective echoes that of other government representatives, where there is a concern that community groups or development organizations acting on their own initiative may achieve suboptimal results through lack of coordination, standardization and appropriate regulation (Sendanayake, 2016). Multiple organizations operating in different communities with somewhat different messages and programmes may boost uptake in the short term, but pose challenges for maintenance of assets designed to varying technical standards. The same representative added that the general lack of knowledge households and communities have surrounding DRWH needs to be addressed. In water-stressed districts where groundwater potential is low and water quality is poor, the Appropriate Technology Centre promotes DRWH using a 'revolving fund' approach through NGOs such as the Shuuka Development Foundation, Katosi Women Development Trust, Busoga Trust and Ugandan Muslim Development Association (Ministry of Water and Environment, 2017). Therefore, much of the sensitization, mobilization and financing work is undertaken by these locally based NGOs, which may explain in part why membership numbers are so high in these regions, and why membership is a strong predictor of DRWH uptake. However, a March 2016 law to curb political electioneering by NGOs in the country may limit the ability of NGOs and other community-based organizations to support DRWH if this is seen as implying criticism of the government or its policies.

Income and finance mechanisms

It was difficult to clearly determine whether monthly income usefully discriminates between adopters and non-adopters due to the skewed (towards poorer households) nature of the sample population. However, analyzing the income of adopters, without comparing them to non-adopters, provides some interesting insights. Although irregular monthly income was a common barrier to adoption noted by respondents, an approximate average monthly income was given for 96% of adopters. The modal average (noted by 37% of respondents) was between US$ 30 and US$ 59 per month, followed by less than US$ 30 per month at 25%, with only 12%

registering more than US$ 89 per month. Even with corrections for 2011 purchasing power parity of US$ 1.90 a day, 62% or respondents are still at or below the national poverty line (Ministry of Finance, 2014; Uganda Bureau of Statistics, 2014a, 2014b). Income is not, by itself, a good predictor of propensity to adopt.

Focus group discussions suggested that stakeholders were concerned that households perceive alternative water sources as 'free' and so are often not incentivized to pay (up to 40% of) the initial cost of DRWH. The usual model for installation involves a modest co-pay plus labour input into the construction of DRWH systems, with the community or intermediary organization picking up the rest of the cost. In some cases (HIV widows, extreme poverty) the local organization may agree to pay 100% of capital costs. This suggests that *absolute* income levels are not a strong predictor of uptake, in part because the surveyed population may have been too poor for the study to detect any relation between household income and likelihood of implementation (i.e., no price elasticity of demand) but also because many are opposed to the idea that water should cost anything. Studies elsewhere have also found that DRWH systems are often more attractive to middle-income households, who are more accustomed to consumerist logic (Ahmed et al., 2013).

Often, to qualify for assistance from community or intermediary organizations in setting up DRWH infrastructure, families must demonstrate that they have improved sanitation facilities. This poses several problems for would-be water harvesters, especially in low-income families, as the average cost of improved sanitation is approximately US$ 50, and the community or intermediary organizations that support sanitation development are often different from the ones promoting DRWH.[4] Again, it may be that lack of integration and coordination with the water, sanitation and hygiene (WASH) sector is unduly suppressing progress in water supply and sanitation services. With 62% of adopters at or below the national poverty line, there is great potential for this practice to exclude those most at risk of household water insecurity. It might be assumed, therefore, that access to credit would be necessary for gaining improved sanitation. In fact, though in most cases with an average income less than US$ 59, 80% *did* claim to have access to credit. Consistent with previous studies (Ahmed et al., 2013; Baguma & Loiskandl, 2010), this research found access to credit to be moderately positively correlated with the adoption of DRWH ($r^2 = 0.490$, $n = 66$, $p \leq .001$). Attitudes to using credit for DRWH were also fairly positive, especially where the credit organization was locally embedded and well known, again pointing to the important role played by intermediary NGOs.

However, owing to the absence of a well-developed formal credit system for water and sanitation infrastructure in Uganda, many households tended to use informal, potentially exploitative, credit systems (such as village groups or informal money-lenders with high interest rates) to improve water services. Mpuga (2010) estimates that only 43% of Ugandans have access to a bank account, and fewer than 5% have access to a microfinance scheme. To support poorer households that cannot afford to adopt DRWH, researchers and practitioners have proposed improving access to microfinance for sanitation (Rosenberg, Gonzalez, & Narain, 2009), a finding replicated in our study.[5] Perhaps NGO involvement earlier in the adoption process, supplying hardware subsidies for water supply and sanitation facilities, to reduce dependence on underdeveloped, informal, potentially exploitative credit systems, would be a useful way of improving uptake (Mpuga, 2010; Rosenberg et al., 2009). Reporting on the role of microfinance in increasing access to improved WASH services, Hadi (2000) observed that households involved with credit programmes were more likely to be water secure than others who were equally poor but not involved in such programmes. There may also be room for the development of microfinance (along the lines of the well-known Grameen Bank model) for WASH investment in Ugandan communities (Davis, White, Damodaren, & Thostren, 2008).

Lifecourse dynamics

This study found that age of the head of household is strongly related to the adoption of DRWH, with a significant difference in the mean age of adopters (51 years) versus non-adopters (38 years). The preponderance of older adopters may be linked to several factors, including

- having had more time available to accrue sufficient capital (for example, through having more income-generating members of the household or having been working longer);
- being better able to access to credit, and related knowledge and resources; and
- having more social capital, enabling them to draw on assistance with construction and maintenance (Coleman, 1988; Mpuga, 2010; Seferiadis, Cummings, Zweekhorst, & Bunders, 2015).

There is no indication in our data that attitudes to DRWH vary with age, as our sample of respondents was overwhelmingly positive (90%) towards DRWH (*contra* some reports, e.g., Ahmed et al., 2013), further suggesting that the significance of age is more in the material than in the affective domain.

Interestingly, given the above, we did not find that level of educational attainment was linked to adoption of DRWH. This is consistent with some studies (e.g., Baguma & Loiskandl, 2010) but contradicts others (e.g., Ahmed et al., 2013). The latter study suggests that, as one might expect, more education led to more exposure to information and critical perspectives (e.g., about climate change or well-being), resulting in a more positive view of DRWH and related water security strategies. It is notable that the leadership of older, and better educated, community members (especially clergy and medical staff) can be a critical factor in promoting uptake. 'Capabilities' therefore reside in both individuals and in communities, whose networks may be more or less rich and more or less supportive of a given innovation like DRWH (Jepson et al., 2017). Data on education and DRWH uptake are complicated by the potential for NGOs to act as vehicles for perception change and capacity building, separate from formal educational attainments. It is possible that education becomes less significant in relation to DRWH adoption due to the ability of such organizations to expose households of varied education levels to this information, mobilizing households regardless of education. Perhaps Ahmed et al.'s study communities lacked the organizations present in our Ugandan study areas.

An ironic outcome of the above is that it may be younger households that are less likely to adopt DRWH, as a consequence of a lack of social capital, including knowledge, social networks and access to (micro)finance mechanisms (Coleman, 1988). What's more, ethnic and other sorts of population mixing in communities can pose challenges for knowledge dissemination and the building of social trust across communities, both of which are necessary for growing DRWH uptake (Letki, 2008). Grassroots NGOs, such as local or district WASH programmes, were often created by local churches (especially the Church of Uganda), implying that parishioners would have easier and quicker access to necessary information and other resources to support uptake.

No clear trend between DRWH adoption and gender was found, although slightly higher rates of male household heads were found in non-adopters. Consequently, our findings do not support the widely held assumption that adoption may be higher among females because of their proximity to the practical daily burden of water collection and management. Possibly this is due to the prevalence of male-headed households and the tendency revealed in the questionnaire survey for younger girls (rather than women) to be the main collectors of water. If this is so (and the authors have seen similar results in other parts of Uganda), then there is an important asymmetry here between those who are in a position to make decisions about DRWH adoption (male heads of households) and those who most acutely feel the impact of their decisions (young girls and women). Stevenson et al.'s (2012) study of water insecurity in Ethiopia also found that women's vested interests in household water decisions were often suppressed or obscured by patriarchal control over resource allocation and other household decisions.

As in other studies (Ahmed et al., 2013; Baguma & Loiskandl, 2010; Gould & Nissen-Petersen, 1999), our survey found that family size is correlated reasonably strongly with propensity to adopt DRWH. The larger the family, the more likely adoption becomes (r^2 = 0.490, n = 66, $p \leq .05$). As Figure 5 shows, the family size of adopters is significantly larger than that of non-adopters, with over 75% registering between five and nine family members, compared with an average rural family size country-wide of 4.8. There is some evidence that this correlation may be at least in part a product of explicit NGO policy around DRWH promotion. In an informal interview with a member of the Busoga Trust, it was suggested that the community-based NGO specifically targets households with larger family sizes because the introduction of DRWH

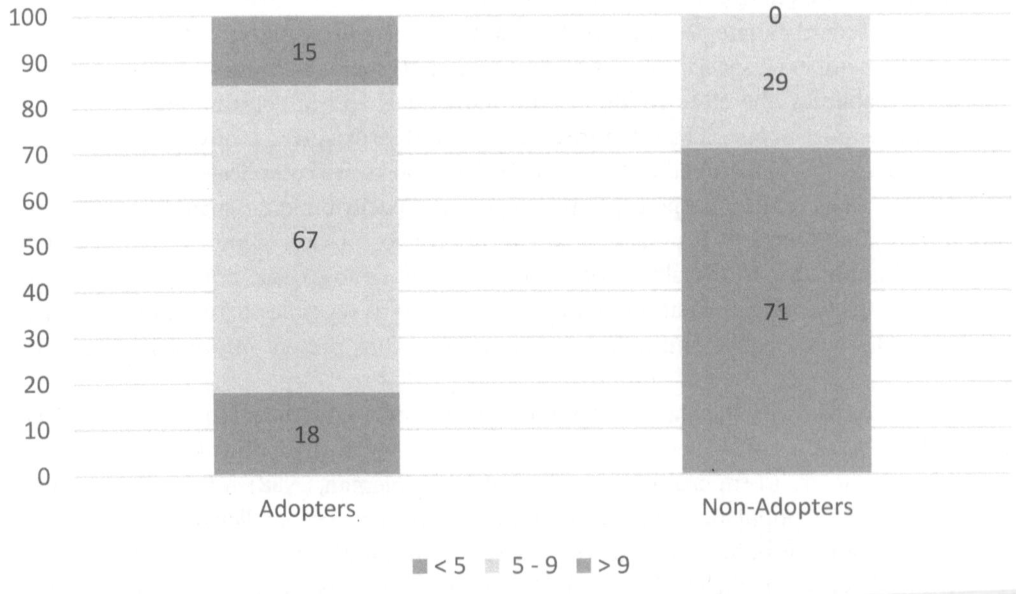

Figure 5. Family size and propensity to adopt rainwater harvesting. Source: the authors.

tanks in these households benefits a larger number of people and may also reach more vulnerable people, particularly young children and the elderly. The authors have found this sort of policy bias, towards benefiting larger families, elsewhere in Uganda as well.

Beyond simple propensity to adopt, there appears to be little relationship between tank size and family size, possibly because the intermediary organizations tend to promote a standard 4 m^3 or 6 m^3 ferro-concrete tank. This is significant because appropriate tank size is crucial to ensure sufficient capacity to get through the dry season (Campisano and Modica, 2012; Kisakye, Akurut, & Van der Bruggen, 2018). An estimation of per capita per day minimum water demand across periods of rainfall deficit is revealing. Typically, Uganda has up to seven months (approximately 215 days) of rainfall deficit spread over one four-month period and one three-month period (Majugu, 2006; McSweeney, New, Lizcano, & Lu, 2010). As defined by WHO (2013), the minimum amount of water required to survive is 7.5 litres per person per day, based on a lactating woman engaged in moderate physical activity in above-average temperatures (DDM, 2012; WHO, 2013). This means that to satisfy only minimum hydration requirements (not including basic hygiene and food preparation), a minimum of 900 litres per person would be required to last out the longer four-month dry period. Thus, a 4 m^3 tank has the capacity (if full and functioning perfectly for the duration of a worst-case drought period) to supply water for only slightly more than four people. With this in mind, it is no surprise that 78% of households with more than five residents indicated they run out of water during dry periods and must revert to alternative, potentially less safe, 'unimproved' water sources.[6] This was confirmed in the data set: 74% of all DRWH adopters surveyed also used alternative water sources, such as open springs, boreholes, vendors and lakes, during periods of deficit, suggesting that DRWH is seen as one of several water sources that can be used in dry periods and not as the sole source on which to depend. Of the three-quarters of adopters who made frequent use of alternative sources, 65% indicated that tanks running out of water was the main reason, while 29% quoted operation and maintenance issues. Possession of a DRWH tank did decrease the distance travelled for alternative water supply, though only slightly, from 2.5 km for non-adopters to 2 km for adopters, still well over the 1 km written into the Millennium Development Goals. The quality of alternative supplies in the study areas was often dubious (Figure 3, left-hand panel).

The above suggests that NGOs involved in promoting DRWH are sometimes promoting a standard-size tank without proper calculation of household size, and therefore need differentials. In some locations, the standard size recommended by the government is 6 m^3, though beneficiaries must contribute higher co-payments to obtain larger sizes, leading to smaller tanks being constructed. In addition to size specification, however, we identify exclusivity of supply (the extent to which a household may or may not have access to other sources) as a key variable in determining the adequacy of a given tank size. Linked to this is the issue of maintenance, since tanks need periodic (at least annual) cleaning to ensure maximum storage volumes and minimization of propensity to store 'bad' water. With regard to operations and maintenance, two of the three NGOs we worked with indicated that after the installation of the tank, their assistance relating to the tank stops, and maintenance becomes the responsibility of the household. With limited resources and skills it may be difficult for households to maintain their tanks, leading to broken pumps or taps, cracks in the tank, and collapsed lids, all of which were seen in the survey. In the focus groups, four district representatives mentioned that DRWH tanks frequently fail to remain operational – in one community of 150 ferro-concrete tanks, 28 were broken 'beyond repair' within only a few years of commissioning. A census of tanks conducted by the authors during the 2017 and 2018 field seasons found that about 10% had suffered (usually preventable) damage inhibiting operation.[7]

Land tenure

A final factor explored in our research and affecting DRWH adoption was land tenure, which had a significant part to play. We found a strong, positive correlation between land ownership and adoption, which was also statistically significant ($r^2 = 0.757$, $n = 65$, $p < .001$). As with other household investments, being free from the worry of eviction appears to embolden households to invest in fixed technical assets like DRWH. Conversely, not owning one's own home or land may deter investment decisions (MWE, 2014; Smucker, Glenn, White, & Bannister, 2000). We also discovered a significant population living on 'disputed land', a situation that sometimes arises in the context of traditional/tribal land tenure systems (where land is given or loaned by patronage) or where there are contestations of inheritance (Figure 6). Interestingly, while a few leaseholders have invested in DRWH or similar innovations, no occupants of 'disputed land' have done so, indicating that security of tenure is an important determinant (Ahmed et al., 2013; Baguma & Loiskandl, 2010).

However, the link between tenure and propensity to adopt may not be as simple as just implied, not least because property owners may be trying to derive multiple benefits from their limited built and land assets. For example, studies elsewhere have shown that property owners are often attempting to manage land for domestic (subsistence) cropping or set aside for future building projects (accommodating future generations, subletting space for

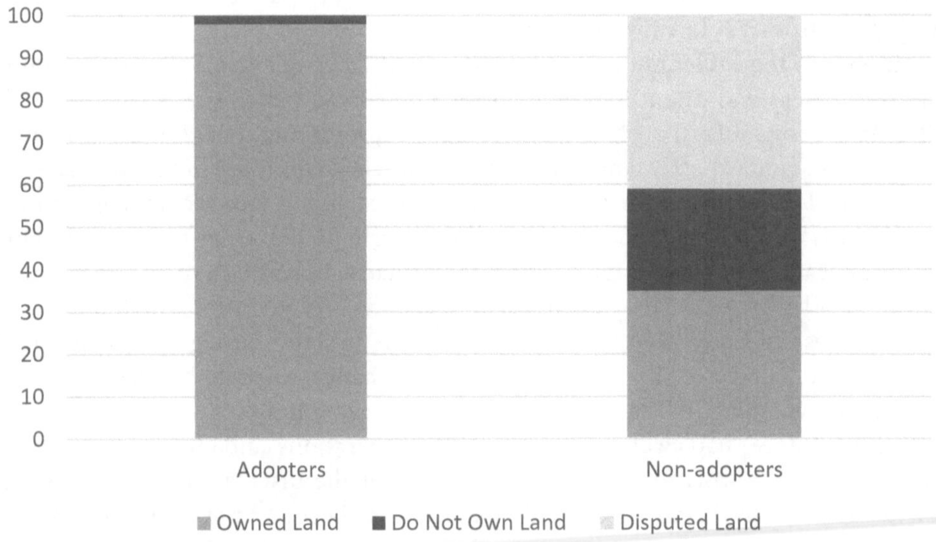

Figure 6. Land tenure and propensity to adopt rainwater harvesting. Source: the authors.

poorer households, etc.). In about a quarter of our study households, there was concern about the space requirements of DRWH. Some householders also mentioned the competing space requirements of sanitation facilities (pit latrines, Ecosan toilets, etc.), which, as we noted earlier, are often prioritized over DRWH or other water supply technologies. And in a few cases adopters clearly articulated that the collected rainwater was not so much for domestic consumptive uses as for the production of market garden produce or a resource for direct sale to nearby households (e.g., for cash to pay school fees) and therefore linked to cash-earning strategies. These findings underscore the importance of understanding what property owners in low-income and high-population-density urban areas want in the context of water services technology innovation and diffusion. Water 'produced' through DRWH can offer many material advantages, some of which are obvious (hydration, hygiene) while some are not (raising cash for school fees, watering animals).

Implications and conclusions

This study of adoption trends of DRWH in three rural Ugandan districts identified both positive and negative factors contributing to the current picture on DRWH adoption. On our evidence we can confidently say that NGOs and community-based organizations (e.g., women's groups and faith groups) are central to the upscaling of DRWH in rural districts in Uganda, but that they can also complicate uptake decisions where they are not sufficiently integrated with other actors in the broader WASH community. For example, it may not be optimal (for district-scale water security) to have different NGOs and community-based organizations focussing on different groups within the broader community, using different finance models and even slightly different physical specifications for the systems themselves. Access to credit and membership in community or intermediary organizations that may offer or organize credit mechanisms (e.g., revolving credit or microfinance schemes) was key to adoption. Therefore, the way these organizations operate, the communities they work with and the ways they implement the technology are all crucial in determining adoption trends. Indeed, this research has highlighted the potential importance of a relational approach to water security and the hydro-social process as the interactions between communities and these organizations largely determines whether households have access to, and are able to adopt, DRWH to improve their water security. Jepson et al.'s (2017) 'capabilities' need, we argue, to be understood as operating in a nested fashion, linking individual household decision makers with the often complex hierarchy of supporting organizations, institutions and policies operating at different spatial scales.

With a greater understanding of adoption trends and drivers, governmental and non-governmental organizations may be able to make more informed decisions on where to focus their work in an attempt to link with communities or households that may previously have been marginalized but that would still benefit from access to DRWH. For example, more attention may be needed to reach those households with small family sizes, with young heads of household, or those who do not own the land on which they reside. We found that all three sub-groups are currently under-represented in the population of those enjoying the benefits of DRWH.

The potential for DRWH to improve household water security, as well as to help with a range of other co-benefits (including gender equality, school attendance, and livelihoods) is high. From a 'capabilities' perspective, DRWH emerges as more than a mere technical 'fix' for household water insecurity; its socio-technical characteristics include links to lifecourse dynamics, governance, regulation, finance and well-being. The fundamental 'capability' that needs development and promotion is the ability to be effectively involved in decisions that affect the ways people and water interact. This article presents some insights into factors affecting adoption, but further multidisciplinary research is required to refine recommendations on policy and practice across a range of stakeholders and actors. Addressing the WASH challenge in the Global South through well-theorized and empirically grounded research should be a matter of global priority.

Notes

1. The World Health Organization defines an 'improved water source' as a type of water source that, by nature of its construction or through active intervention, is likely to be protected from outside contamination, in particular from contamination with faecal matter (WHO/UNICEF, 2015). In the image at right in Figure 2, the 'improved' designation applies because not only is collection structured and made more secure via the retaining wall and tap, but also the uphill area is fenced off to try to prevent contamination.
2. Which is, after all, probably what a 'right to water' ultimately means – see Staddon, Appleby, and Grant (2011).
3. We recognize that there will therefore be a bias in the sample against those households less networked within the community, which would make them less likely to be reached through encountering them when visiting other households.
4. Somewhat akin to the 'community-led total sanitation' approach, it may be useful to think in terms of a 'community-led total water services' approach that tackles water supply and sanitation simultaneously.
5. Lack of an insurance services market (which might, for example, cover DRWH installations) is also a constraint on uptake.
6. Research conducted during the 2018 field season found that tank owners often combined tank water with other sources and even, in some cases, sold tank water for cash income (to pay school fees, for example).
7. Separate research currently underway is examining the root cause pathways to failure in ferrocement rainwater tanks.

Disclosure statement

No potential conflict of interest was reported by the authors.

Funding

This work was supported by Lloyd's Register Foundation, a charitable foundation helping to protect life and property by supporting engineering-related education, public engagement and the application of research.

ORCID

Chad Staddon http://orcid.org/0000-0002-2063-8525

References

Abdulla, F. A., & Al Shareef, A. W. (2009, Retrieved January 2, 2016). Roof rainwater harvesting systems for household water supply in Jordan. *Desalination* [Online], *243*(1), 195–207.

African Ministers Council on Water (AMCOW). (2008). *Fresh water under threat: Africa. Vulnerability assessment of freshwater resources to environmental change.* Nairobi, Kenya: United Nations Environmental Program and Water Research Commission.

Ahmed, A., Mustafa, U., & Khalid, M. (2011). *Impact of Rooftop Rain Water Harvesting (RRWH) technology on women time allocation in Hilly and Fragile Areas of Bagh and Battagram Districts.* CEECC Working Papers. (1). Islamabad. [online].

Ahmed, I., Onwonga, N., Mburu, M., & Elhadi, A. (2013, Retrieved January 3,2016). Evaluation of types and factors influencing adoption of rainwater harvesting techniques in Yatta District, Kenya. *International Journal of Education and Research* [Online], *1*(6), 1–14.

Albaum, G. (1997). The likert scale revisited. *International Journal of Market Research*, *39*(2), 1–21.

Allan, T. (2011). *Virtual water: Tackling the threat to our planet's most precious resource.* IB Taurus.

Ariyabandu, R. D. S. (2001). *Household water security using rainwater harvesting.* Paper presented at Domestic Roofwater Harvesting Workshop, Sri Lanka.

Baguma, D., & Loiskandl, W. (2010). Rainwater harvesting technologies and practises in rural Uganda: A case study. *Mitigation and Adaptation Strategies for Global Change*, *15*(4), 355–369.

Bakker, K. (2012). Water security: Research challenges and opportunities. *Science*, *337*, 914–915.

Campisano, A., & Modica, C. (2012). Optimal sizing of storage tanks for domestic rainwater harvesting in Sicily. *Resources Conservation and Recycling*, *63*, 9–16.

Coleman, J. S. (1988). Social capital in the creation of human capital. *American Journal of Sociology*, *94*, 95–120.

Cresswell, J. W. (2002). *Research design: Qualitative, quantitative, and mixed methods approaches.* London: Sage Publications.

Danert, K., & Motts, N. (2009). *Domestic rainwater harvesting and the water sector in Uganda.* Washington, DC: Sub-Sector Analysis enterpriseWorks/VITA.

Danielson, S., Tuler, S. P., Santos, S. L., Webler, T., & Chess, C. (2012). Three tools for evaluating participation: Focus groups, Q method, and surveys. *Environmental Practice*, *14*, 101–110.

Davis, J., White, G., Damodaren, S., & Thostren, R. (2008). Improving access to water supply and sanitation in urban India: Microfinance for water and sanitation infrastructure development. *Water Science and Technology*, *58*(4), 887–891.

Department of Disaster Management (DDM). (2012). Integrated rainfall variation impacts, needs assessment and drought risk management strategy. *Global Facility for Disaster Reduction and strategy* [online]. Retrieved from http://gfdrr.org/sites/gfdrr/files/UGANDA_PDNA_Report_2012.pdf

Flick, U. (2014). *An introduction to qualitative research.* London: Sage.

Flynn, A., Tremblay, P. F., RFehm, J., & Wells, S. (2013). A modified random walk door-to-door recruitment strategy for collecting social and biological data relating to mental health, substance use, addiction, and violence problems in a Canadian community. *The International Journal of Alcohol and Drug Research*, *2*(2), 7–16.

Forouzanfar, M. H., Alexander, L., Anderson, H. R., Bachman, V. F., Biryukov, S., Brauer, M., ... Delwiche, K. (2015). Global, regional, and national comparative risk assessment of 79 behavioural, environmental and occupational, and metabolic risks or clusters of risks in 188 countries, 1990–2013: A systematic analysis for the global burden of disease study 2013. *Lancet*, *386*(10010), 2287–2323.

Funk, G., Senay, A., Asfaw, J., Verdin, J., Rowland, J., Michaelson, G., ... Choularton, R. (2005). *Recent drought tendencies in Ethiopia and equatorial-subtropical eastern Africa.* Washington, DC: FEWS-NET.

Ghimire, S. R., & Johnston, J. M. (2015). Traditional knowledge of rainwater harvesting compared to five modern case studies. *World environmental and water resources congress 2015: Floods, droughts, and ecosystems*, 182–193. doi:10.1061/9780784479162.017

Gillham, B. (2008). *Developing A Questionnaire* (2nd ed.). London: Continuum.

Gould, J., & Nissen-Petersen, E. (1999). *Rainwater catchment systems for domestic supply*. Rugby, UK: Intermediate Technology Publications, UK.

Hadi, A. (2000). A participatory approach to sanitation: Experience of Bangladeshi NGOs. *Health Policy and Planning, 15*(3), 332–337.

Jepson, W., Budds, J., Eichelberger, L., Harris, L., Norman, E., O'Reilly, K., ... Young, S. (2017). Advancing human capabilities for water security: A relational approach. *Water Security, 1*, 46–52. ISSN 2468-3124.

Kafeero, S. (2018, January 21). Richest, Poorest Districts. *Daily Monitor*. Retrieved from www.monitor.co.ug/News/National/Richest-poorest-districts-uganda/688334-4271356-wp6ug/index.html

Kahinda, J. M., Taigbenu, A. E., & Boroto, J. R. (2007). Domestic rainwater harvesting to improve water supply in rural South Africa. *Physics and Chemistry of the Earth, Parts A/B/C, 32*(15–18), 1050–1057.

Kisakye, V., Akurut, M., & Van der Bruggen, B. (2018). Effect of climate change on reliability of rainwater harvesting systems for Kabarole District, Uganda. *Water Journal, 10*(71). doi:10.3390/w10010071

Lanka Rainwater Harvesting Forum. (2001). *Domestic roof water harvesting and water security in the Humid Tropics*. Report Number: ERB IC18 CT98 0276. Retrieved September 9, 2015, from www2.warwick.ac.uk/fac/sci/eng/research/civil/crg/dtu-old/pubs/reviewed/rwh/eu/d5.pdf

Letki, N. (2008). Does diversity erode social cohesion? Social capital and race in British neighbourhoods. *Political Studies, 56*, 99–126.

Linton, J. (2010). *What is water?: The history of a modern abstraction*. Vancouver, BC: UBC Press.

Lovelle, M. (2016). Co-operation and the Brahmaputra: China and India water sharing. *Future Directions International*. Retrieved from http://www.futuredirections.org.au/publication/co-operation-and-the-brahmaputra-china-and-india-water-sharing/

Majugu, A. (2006). *Investigation of the scientific basis and reliability of observed local indicators for rainfall onset and climatic variability and trends over the Rakai-Kibanda area* (Unpublished report). Ugandan Department Meteorology, Kampala.

Mati, B. M. (2007). 100 ways to manage water for small hold, agriculture in Eastern and Southern Africa, a compendium of technologies and practices. *SWMnet Working Paper 13, IMAWESA*.

McSweeney, C., New, M., Lizcano, G., & Lu, X. (2010). The UNDP climate change country profiles improving the accessibility of observed and projected climate information for studies of climate change in developing countries. *American Meteorological Society, 91*, 157–166.

Ministry of Finance, Planning and Economic Development. (2014). *Uganda poverty status report*. Kampala: Economic Development Policy and Research Department, Government of Uganda.

Ministry of Water and Environment. (2010). *Self supply experiences in Uganda*. Kampala: Government of Uganda.

Ministry of Water and Environment. (2014). *Water and environment sector performance report*. Kampala: Government of Uganda.

Ministry of Water and Environment. (2017). *Water and environment sector performance report*. Kampala: Government of Uganda.

Mpuga, P. (2010). Constraints in access to and demand for rural credit: Evidence from Uganda. *African Development Review, 22*(1), 115–148.

Nayebare, S. R., Wilson, L. R., Carpenter, D. O., Dziewulski, D. M., & Kannan, K. (2014). A review of potable water accessibility and sustainability issues in developing countries–Case study of Uganda. *Reviews of Environmental Health, 29*(4), 363–378.

Nsubuga, F. N., Namutebi, E. N., & Nsubuga-Ssenfuma, M. (2014). Water resources of Uganda: An assessment and review. *Journal of Water Resource and Protection, 6*(14), 1297.

Owen, G., & Goldin, J. (2015, December 12). Assessing the relationship between youth capabilities and food security: A case study of a rainwater harvesting project in South Africa. *Water SA* [Online], *41*(4), 541–548.

Pullan, R. L., Freeman, M. C., Gething, P. W., & Brooker, S. J. (2014). Geographical inequalities in use of improved drinking water supply and sanitation across sub-Saharan Africa: Mapping and spatial analysis of cross-sectional survey data. *PLoS Medicine*, *11*(4), 1–17.

Rogerson, P. (2010). *Statistical methods for geography: A student's guide* (3rd ed.). London: Sage Publications.

Rosenberg, R., Gonzalez, A., & Narain, S. (2009). The new moneylenders: Are the poor being exploited by high microcredit interest rates? *Consultative Group to Assist the Poor* [Online], *1* (15). Washington. Retrieved March 28, 2016.

Scott, C. A., Meza, F. J., Varady, R. G., Tiessen, H., McEvoy, J., Garfin, G. M., … Pineda-Pablos, N. (2013). Water security and adaptive management in the Arid Americas. *Annals of the Association of American Geographers*, *103*, 280–289.

Seferiadis, A. A., Cummings, S., Zweekhorst, M. B., & Bunders, J. F. (2015, Retrieved March 13, 2016). Producing social capital as a development strategy: Implications at the micro-level. *Progress in Development Studies* [Online], *15*(2), 170.

Sendanayake, S. (2016). Life cycle analysis of ferro-cement rainwater tanks in Sri Lanka: A comparison with RRC and HDPE tanks. *International Journal of Applied Engineering Research*, *12*(2), 1–8.

Sin, H. C. (2003). Interviewing in 'place': The socio-spatial construction of interview data. *Area*, *35*(3), 305–312.

Siwar, C., & Ahmed, F. (2014, Retrieved November 7, 2015). Concepts, dimensions and elements of water security. *Pakistan Journal of Nutrition* [Online], *13*(5), 281–286.

Smet, J. (2003, March). Well fact sheet: Domestic rainwater harvesting. *Blog posted*. Retrieved from www.lboro.ac.uk/orgs/well/resources/fact-sheets/fact-sheets-htm/drh.htm

Smucker, T., Glenn, R., White, A., & Bannister, M. (2000). Land tenure and the adoption of agricultural technology in Haiti. *International Food Policy Research Institute* [Online]. Working Paper 6. Retrieved March 28, 2016.

Staddon, C. (2010). *Managing Europe's water resources: 21st century challenges*. Ashgate Press.

Staddon, C., Appleby, T., & Grant, E. (2011). A right to water - A geographico-legal perspective. In F. Sultana & A. Loftus (eds), *The right to water: Politics, governance and social struggles* (1st ed., pp. 61–77). London, UK: Earthscan.

Staddon, C., & Genchev, S. (2013). Geographical and temporal dimensions of Bulgarian attitudes towards the rural environment. *Czech Sociological Review* [Online], *49*(3), 375–401.

Staddon, C., & James, N. (2014). Water security: A genealogy of emerging discourses. In G. Schneier-Madanes (ed), *Globalised water: A question of governance* (pp. 261–276). Dordrecht, The Netherlands: Springer.

Stevenson, E. G., Greene, L. E., Maes, K. C., Ambelu, A., Tesfaye, Y. A., Rheingans, R., & Hadley, C. (2012). Water insecurity in 3 dimensions: An anthropological perspective on water and women's psychosocial distress in Ethiopia. *Social Sciences Medica*, *75*(2), 392–400.

Sun, F., Staddon, C., & Chen, M. (2016). Developing and applying water security metrics in china: Experience and challenges. *Current Opinion in Environmental Sustainability*, *21*, 29–36.

Terry, A., McLaughlin, O., & Kazooba, F. (2015). Improving the effectiveness of Ugandan water user committees. *Development in Practice*, *25*(5), 715–727.

Thayil-Blanchard, J., & Mihelcic, J. (2015). *Assessing domestic rainwater harvesting storage cost and geographic availability in Uganda's Rakai District*. In F. Memon and S. Ward (Eds.), *Alternative water supply systems* (pp. 131-152). London: IWA Publishing.

Turner, S. (1999). *Design of storage tanks for rainwater harvesting, Indian Institute of Technology*. Delhi, India: Indian Institute of Technology.

Uganda Bureau of Statistics. (2014a). *Statistical Abstract*. Retrieved from www.ubos.org/online files/uploads/ubos/statistical_abstracts/Statistical_Abstract_2014.pdf

Uganda Bureau of Statistics. (2014b). *National population and housing census* (pp. 19–65). Uganda.

Uganda Bureau of Statistics. (2017a). *The national population and housing census 2014 – area specific profile series: Luwero District, Kampala, Uganda.*

Uganda Bureau of Statistics. (2017b). *The National Population and Housing Census 2014 – Area Specific Profile Series: Mukono District, Kampala, Uganda.*

Uganda Bureau of Statistics. (2017c). *The national population and housing census 2014 – area specific profile series: Wakiso District, Kampala, Uganda.*

UNDP (2018). *Goal 6: Clean water and sanitation.* Retrieved from http://www.undp.org/content/undp/en/home/sustainable-development-goals/goal-6-clean-water-and-sanitation.htm

United Nations Framework Convention on Climate Change (UNFCC). (2007). *Climate change: Impacts, vulnerabilities and adaptation in developing countries.* Retrieved from https://unfccc.int/resource/docs/publications/impacts.pdf

UN-Water. (2015). *United nations world water development report* [online]. Paris: UNESCO. Retrieved from http://unesdoc.unesco.org/images/0023/002318/231823E.pdf

Valentine, G. (1999). Doing household research: Interviewing couples together and apart. *Area, 31*(1), 67–74.

Webb, P., & Iskandraani, M. (1998). Water insecurity and the poor. Issues and research needs. *Center for Development Research* [Online], *2*, 66. Retrieved March 10, 2016.

WHO. (2017). *2.1 billion people lack safe drinking water at home, more than twice as many lack safe sanitation* [online]. Retrieved February 22, 2018, from www.who.int/mediacentre/news/releases/2017/water-sanitation-hygiene/en

WHO/UNICEF Joint Monitoring Programme. (2015). *Progress on sanitation and drinking water: 2015 update and MDG assessment.*

World Health Organisation. (2013). *Technical notes on drinking water, sanitation and hygiene practices.* Geneva: WHO.

Wutich, A., Budds, J., Eichelberger, L., Harris, L., Norman, E., O'Reilly, K., . . . Young, S. (2017). Advancing methods for research on household water insecurity: Studying entitlements and capabilities, socio-cultural dynamics, and political processes, institutions and governance. *Water Security, 2*, 1–10.

WWAP (United Nations World Water Assessment Programme). (2018). *The united nations world water development report 2018: Nature-based solutions* [online]. Paris: UNESCO. Retrieved August 21, 2018, from http://unesdoc.unesco.org/images/0026/002614/261424e.pdf

Zingiro, A., Okello, J. J., & Guthiga, P. M. (2014, Retrieved August 26, 2015). Assessment of adoption and impact of rainwater harvesting technologies on rural farm household income: The case of rainwater harvesting ponds in Rwanda. *Environment, Development and Sustainability, 16*(6), 1281–1298.

Water security and the pursuit of food, energy, and earth systems resilience

Christopher A. Scott, Tamee R. Albrecht, Rafael Routson De Grenade, Adriana Zuniga-Teran, Robert G. Varady and Bhuwan Thapa

ABSTRACT

This article addresses the emergence and interrelation of food, energy, and water security in terms of resource use and the ensuing societal and environmental outcomes. For decades, food security and energy security have been well-accepted, operational concepts. Water security is the latest entrant, yet the implications of water insecurity for food, energy and earth systems resilience have not been adequately considered. This article examines how and why this is so – particularly with growing water scarcity and insecurity that may compete with energy and food security – and emphasizes the critical need to link water-energy-food nexus approaches to earth systems resilience.

Introduction: a brief review of food, energy and water security

In the past century, two devastating world wars, accumulating population growth, rising urbanization, accelerated climate change, and a host of other mounting problems have affected human security worldwide. Meeting these security concerns requires adequate amounts of and access to critical resources such as food, water and energy. By 2050, for instance, UN Water (2015) projects that global food production will increase about 60%, which will require 45% more energy and 30% more water.

The interconnections between physio-ecological processes and the human dimensions of security were recognized by the United Nations Development Programme in the mid-1990s (Scott & Thapa, 2015). Food production, for instance, is responsible for significant changes in freshwater supplies (Ingram, Ericksen, & Liverman, 2010), while energy security is fundamental to food and water security because these sectors increasingly rely on uninterrupted energy supply. Further, water is needed to produce food; current production practices and irrigation technology make the agricultural sector the world's largest water user. In extreme instances, competition for resources – especially non-renewable ones – has led to open conflict. Cooperative resource-sharing can be a channel for international peace-building that can enhance human security (Theisen, 2008).

The current conception of the water-energy-food (WEF) nexus was developed in the early 2000s, in response to a worldwide food and economic crisis. The nexus refers to the interrelationships and mutual interdependence of water resources for food and energy production; energy requirements for water management, agriculture, and food processing and transportation; and food trade-offs associated with alternate water or energy allocations (Hoff, 2011; Scott, Kurian, & Wescoat, 2015). Numerous international conferences have recognized the WEF (or variously, FEW) resource nexus, and some national governments – chiefly Germany – actively promote the framework. Academics have conducted relevant research on the nexus concept; nonetheless, efforts to put it into practice using specific tools supported by appropriate and consistent methodologies are limited (Albrecht et al., 2018).

Some actors may prioritize concerns over resource security – the availability of sufficient quantity and quality for present and future human and environmental needs – of either water, energy, or food. But viewing the entirety of human security through the WEF nexus is distinctive because it aims to consider the three sectors as equally important (Benson, Gain, & Rouillard, 2015). Only recently have the security aspects of the nexus begun to gain the attention they deserve (Benson et al., 2015; Kumar, Bassi, Narayanamoorthy, & Sivamohan, 2014). Cross-sectoral analyses pivoted on the nexus elements offer a fresh way to approach security – both human and environmental.

Advancing the nexus concept, de Grenade et al. (2016) suggested a need to bring earth system resilience as a dynamic and fundamental conceptual framing into nexus research and practice. Complex, dynamic, interlinked earth systems provide resources, are the source of multiple drivers and stressors, and are affected by, but exist independent of, human actions. Furthermore, resilience brings agility to the conceptual approach – one that encourages adaptive action and focuses on the dynamic capacity of social and ecological systems. This resilience understanding goes beyond the static equilibrium and limits-to-growth conceptualizations of much sustainability thinking that stem from the 1987 report of the Brundtland Commission and, earlier, the 1972 *Limits to Growth*. The earth systems approach advanced by de Grenade et al. (2016) places particular emphasis on the ecological, hydrological and climate system dynamics that underpin societal use of water, energy and food resources. Earth systems are complex, interrelated, and difficult to understand in isolation, much less in context. When human-driven forces cause changes in not one aspect but multiple phases of the system simultaneously, the resulting changes may prove beyond the limits of adaptation for many living species. Earth system resilience considers geologic, biologic and atmospheric processes in relation to human adaptation.

This article is organized on the conceptual approach of interdependent resources and intertwined securities. We trace the emergence of food, energy, and subsequently, water security constructs – each with a historical and institutional logic. Food-security concerns surfaced early, particularly in the second half of the twentieth century, in response to civil strife and political instability, uneven income distribution, population growth, lack of credit and capital, distribution and storage problems, and perhaps most significantly, weak institutions and ineffective governance. Interest in energy security emerged on the international stage in the 1970s, prompted by increasingly globalized markets, heightened demand for energy, excessive reliance on fossil fuels, uneven distribution of energy resources, lack of capital, mismanagement of oil and gas supplies, and perhaps most notably, cartelist supply of petroleum. Water security was the latest entrant, emerging now in the early twenty-first century as a clearly recognized global resource-security challenge, even though water scarcity and competition, quality degradation and flooding are centuries-old challenges. Nonetheless, the pursuit of water security is fundamental to the Sustainable Development Goals, the Sendai Framework for Disaster Risk Reduction, and programmes of the World Economic Forum, among other global initiatives.

We consider situations in which water security has unavoidable and inextricably linked consequences for energy and food security and earth systems. We use a nexus-based approach, exploring why, in practice, the nexus remains a water-centric framework and how this may affect possible outcomes. Next, we show how a nexus-based approach may offer unique insights for human security. In the subsequent sections, we present and discuss how a water-security conception of the nexus can be linked to earth system resilience. We also consider the broader resource-security framework and its implications for the earth system, which intrinsically supports human security concerns. We close with reflections on the attainability or elusiveness of water security. Because it is essential for policy recommendations to be context-specific in order to be relevant and actionable, we refrain from drawing out specific policy implications. We do nevertheless raise questions for policy research and applied decision-making.

Evolution of the food-security concept

The notion of food security emerged after World War II; access to food – or food security – was the chief concern of the postwar international humanitarian agenda. In 1945, based on previous League of Nations initiatives, the UN established the Food and Agriculture Organization (FAO), where food security became an international humanitarian goal and a national-planning priority. Sir John Boyd Orr, FAO's first director-general, proposed (unsuccessfully) a World Food Board to end hunger (Salvatici, 2015; Shaw, 2007). In 1957, FAO launched a public-awareness campaign that spawned the World Food Programme to deal with food crises. The first World Food Conference (1974) adopted 20 resolutions to eradicate hunger, and two years later, a new FAO director-general helped activate the concept of 'food security' (access to food by the poor) and established a World Food Council to coordinate efforts (Shaw, 2007).

Other UN agencies joined the effort. In 1976, the International Labour Organization coined the term 'basic needs'; a 1986 World Bank study distinguished between *transitory* and *chronic* food insecurity; in 1988 the International Food Policy Research Institute analyzed the effects of food subsidies in developing countries; and in 1998 Amartya Sen's introduction of 'food entitlement' earned him the Nobel Prize in economics (Shaw, 2007).

International food-security conferences and other dialogues have added such new perspectives as environment, children, water resources, climate change, human rights, nutrition, population, global hunger, population, social development, women, and trade. The World Food Summit (1996) defined food security as 'physical and economic access to sufficient, safe and nutritious food … for all people, at all times'. At the Millennium Summit

(2000), world leaders committed to halve the portion of the world's population that suffered hunger (Salvatici, 2015; Shaw, 2007). And 15 years later, 190 nations agreed to 17 Sustainable Development Goals, the second of which is to end hunger.

Food production is rising, so food security is primarily an equity concern. Paradoxically, death from malnutrition prevails in some developing countries, while obesity shortens lives in developed countries (Shaw, 2007). As global food production has become a business of scale, *waste* – when food is lost either before reaching consumers or is wasted by them – has been an emerging food-security challenge. Also, rising food prices can threaten human security and political stability via protests and riots. In 2007–2008, 48 countries experienced record-high food prices and subsequent social disturbance, which in turn led to further food-price hikes. Governments are key to controlling high food prices, while international organizations and NGOs can assist fragile states (Godfray & Robinson, 2015).

Food security requires equitable distribution, access, and prices; but disagreement on globalization, trade policies, climate change, genetically modified crops, fertilizer availability, and – key for water-security concerns – access to irrigation water impede equity (Scott & Thapa, 2015; Shaw, 2007). Smallholders and rural populations are especially vulnerable to food-security impacts from climate change (Nelson et al., 2016). Other past and ongoing drivers of food insecurity include weak governance, income inequality, inadequate distribution and marketing networks, insufficient monitoring of incipient droughts, lack of stable rural credit mechanisms, and at the extreme, civil unrest.

Although the notion of food security was formalized in the twentieth century, food-security concerns certainly existed before, rising and declining in successive waves according to political and environmental conditions. In the late eighteenth century, starvation was considered inevitable by many, yet the subsequent Industrial Revolution and vast expansion of cultivable land raised production, and in many cases prevented starvation. In the mid-twentieth century, food-security concerns re-emerged with population growth and a collective awareness of limited resources. But Green Revolution productivity increases and birth-rate reductions helped prevent famines (Godfray & Robinson, 2015).

Now the world is experiencing a third wave of food-insecurity challenges, triggered by a food-price crisis and worries about complex dependencies of food security on dynamic earth system processes – rising temperatures, droughts, coastal inundations, and soil depletion and degradation. These concerns are sharpened by continuing population growth, higher per capita consumption, changes in dietary preferences in developing and developed nations alike (chiefly, increased consumption of meat, with its significantly higher water and energy footprints compared to grains and vegetables), increase of waste, and growing limitation of arable land, as well as climate-change-induced shifts in production zones and variability in agricultural productivity. Sufficient resources and capabilities exist to end hunger globally, yet over a billion people (15% of the world's population) continue to lack access to sufficient food (Ingram et al., 2010). In their attempts to achieve food security, policy makers face complex, politically fraught decisions exacerbated by insufficient access to financial resources (Godfray & Robinson, 2015; Kuyper & Struik, 2014; Westengen & Banik, 2016). These challenges are not unique to food security and can limit the pursuit of energy and water security, too.

The challenge of energy security

Since the Industrial Revolution, economies and societies have depended highly on fossil-fuel energy, whose sources are concentrated in certain geographic regions. As a result, energy is arguably the resource most immediate to national-security interests (Scott & Thapa, 2015). The idea of energy security emerged in the late nineteenth century, with the rise of coal-powered warships (Sovacool & Brown, 2010). By World War I, the world's large navies were shifting from domestically produced coal to imported petroleum, rendering them vulnerable to supply interruptions caused by attacks on oilfields, transportation lines, and refineries (Cherp & Jewell, 2011).

After World War II – because of the rise of consumer automobiles and the need to generate electricity for heating, manufacturing, food production and health care – addiction to oil grew markedly. After the end of colonialism, newly independent nations began exporting petroleum, sometimes as a chief source of revenue. This practice poses a risk to political stability and generates energy security concerns in the countries dependent on imports (Cherp & Jewell, 2011; O'Rourke & Connolly, 2003). In 1973, an oil embargo by OPEC (Organization of Petroleum Exporting Countries) members dominated by Arabian countries, caused massive price increases. The 1979 Iranian revolution brought additional price rises, further elevating energy-security concerns. The 1990 Iraqi invasion of Kuwait caused another energy-price crisis. Since then, OPEC members' power to influence prices has diminished (Sovacool, 2007), while over the same period, newly exploited sources of petroleum accompanied by a shift to other energy sources such as natural gas and nuclear energy has reduced costs (Cherp & Jewell, 2011).

Much of the twentieth-century discourse on energy security had been framed by geopolitical theories and strategic-security studies. The recognition of resource limitation, however, along with new modes of analysis, enabled reconceptualization of the problem (Pasqualetti & Sovacool, 2012). Just before the 1973 crisis, the iconic book *Limits to Growth* (Meadows, Meadows, Randers, & Behrens, 1972) propounded the notion of resource scarcity. Although that crisis was not caused by resource scarcity per se, a fear of limits prevailed ('peak oil theory'). In recent years, a more nuanced concept of energy security relies on complex-systems analyses that highlight system interactions, vulnerabilities and resilience (Cherp & Jewell, 2011; Pachauri & Cherp, 2011). Current concerns centre on transitioning from carbon-based energy sources to renewable sources to limit greenhouse gas emissions (McGlade & Ekins, 2015).

In the United States, energy security focuses on system vulnerability – particularly massive electrical-system or power-generation failures. Since the 1950s, energy-security concerns – especially generation and transmission – have highlighted short-term disruptions by technical failures, extreme natural events, even sabotage and terrorism – all attributable to vulnerabilities of infrastructural systems that are becoming more complex. Such concerns were reinforced by major nuclear accidents at Three Mile Island (1979), Chernobyl (1986), and Fukushima (2011), which collectively shrank interest in and support for nuclear energy worldwide. In short, views of energy security shifted somewhat, from purely geopolitical concerns (*Who controls energy supplies?*) to vulnerabilities in energy systems (*How vulnerable are the systems to different disruptions?*) (Cherp & Jewell, 2011).

Neoliberal ideas influenced energy security in the late twentieth century. In the developed world in the 1980s and 1990s, energy supply was sometimes deregulated in favour of markets that would promote investment in energy infrastructure, ensure supply through market diversity, and lead to greater profits for utilities and investors, even while offering cheaper supply – all while making the system more resilient to political disruptions. Market-design flaws and insufficient regulation, however, inhibited market-based strategies to complete energy security (Cherp & Jewell, 2011).

Meanwhile, beginning in the 1970s, minimizing the risk of energy-system failures by diversifying energy options – known as portfolio management – was added to the suite of energy-security strategies. This approach, which considered the probability of various threats, as well as resilience and flexibility, became part of the complex-systems-analysis toolbox. Currently, state-of-the-art evaluation of energy security often uses diversity indices. But such techniques are not universally employed; some nations and regions fail to incorporate diversity and continue to rely on one type of energy source, e.g., natural gas in Eurasia and petroleum in Saudi Arabia.

Contemporary energy-security challenges are being approached from an integrative perspective, as opposed to studying individual components separately (geographic location of sources, type of fuel, distribution mechanism, sector). For example, replacing fossil fuels by renewable-energy sources requires changes in energy infrastructure (electric grids), market incentives (lower prices), design of vehicles (electric cars), the relationship with food production (biofuels), and requirements for water (e.g., more hydroelectric facilities, or the need for more water for fracking). There are also major energy-security implications of climate change adaptation (lower carbon sources) and adaptation impacts (reduced spatial and temporal water availability). Challenges for each component have to be solved urgently, simultaneously, and in an integrated manner.

As with food security, equity concerns are an important challenge for achieving energy security. Integration of scales has been identified as crucial for social equity. Energy security, mainly a national-level concern, needs to be integrated with energy access, mostly a concern at the household or sectoral level (e.g., urban vs. agricultural supply). As the world transitions to alternative sources, the true costs (environmental and social impacts) will become internalized, and energy will be less affordable. This invariably affects developing and energy-poor nations, and vulnerable groups, particularly in poor countries. Disadvantaged populations are especially susceptible to price changes and other energy-system disruptions. In this regard, women in developing societies play a critical role; they are often the end users and are charged with finding fuel. And because alternative-energy-production sites can be easily disrupted – and for cities, often require long-distance transmission – those societies aiming to transition to such sources are more vulnerable to threats. Financial systems also require innovation for energy to reach poor communities (Pachauri & Cherp, 2011).

Finally, we note that climate change – a potent earth system process – has a strong potential to affect energy availability and therefore energy security. Extended droughts and rising temperatures can reduce snowpack, which in turn reduces the volume of water available for hydropower. Less directly, but as importantly, high temperatures raise demand for power such as for air conditioning and cooling systems, especially in urban settings. And climate-change-induced extreme heat and aridity, as well as tropical storms and flooding, can disrupt electric transmission, pipeline transport, and vehicular conveyance of energy materials. At the opposite extreme, some areas may experience severe flooding or coastal sea level rises; these phenomena also can require surges of energy use and are capable of disrupting grids and provisioning networks.

In sum, new, flexible, and innovative governance systems will be needed to address energy-security and energy-access challenges under complex and increasingly uncertain conditions. As noted above, however, the need for sound governance applies not just to energy but also to water and food security.

Whence water security?

Water security is the most recent element of the WEF nexus to emerge as an organizing principle on the global agenda. In the twentieth century, the term 'water security' was first used to communicate concerns regarding water resources, including continuous and equitable access to safe water (Gerlak et al., 2018). Although state-security discourses in the 1940s and 1950s focused on military and other means to control water and other strategic resources, one of the earliest expressions of water security came from Strong (1956), who contextualized it as having 'productive access to water under conditions of scarcity and uncertainty' (p. 6) (Scott & Thapa, 2015). However, it was not until the 2000s that the concept emerged as a paradigm (Cook & Bakker, 2012); and it was first defined by the Global Water Partnership (GWP, 2000, p. 12) as a goal in which 'every person has access to enough safe water at affordable cost to lead a clean, healthy and productive life, while ensuring that the natural environment is protected and enhanced'. Since then, use of the term has grown markedly and captured the attention of numerous international organizations, e.g., Asia-Pacific Water Forum, GWP, UNEP, UNESCO and UN Water (Cook & Bakker, 2012; Gerlak et al., 2018; Varady, Zuniga-Teran, Garfin, Martin, & Vicuña, 2016).

Water-security concerns also reached the international agenda in the early 2000s. In September 2000, the UN adopted the Millennium Development Goals, committing to reduce poverty by 2015. The seventh of these eight goals, to 'ensure environmental sustainability', includes a specific target for water security: to 'halve, by 2015, the proportion of people without sustainable access to safe drinking water and basic sanitation' (p. xiii) (Sachs, 2012). In 2015, the UN adopted the 17 Sustainable Development Goals (SDGs), including one (no. 6) specifically for water security: to 'ensure access to water and sanitation for all', with a new target of 2030.

Since then, the concept of water security has been adopted as a key water-governance objective by governmental and nongovernmental organizations, with growing interest in alternative measures to achieve water security other than infrastructure development (Zeitoun et al., 2016). Policy makers and academics use the term 'water security' with multiple definitions, framings and analytical approaches, based on a diverse set of disciplines in social and physical sciences, engineering, management and law. Water security is now used frequently as a framework of analysis at diverse scales and with different foci, depending on geographic location. For example, in unstable regions, water-security concerns may include bio-terrorism, while in developing and emerging economies, water-quality concerns may be at the fore of their domestic agenda (Cook & Bakker, 2012). Social conflict – violent and nonviolent – is also increasing, particularly among competing users and sectors (Bakker, 2012).

Perhaps more than food and energy security, water security has an intrinsic duality: water can be considered both as productive (a resource) and as destructive (a hazard) (Grey & Sadoff, 2007), although biohazards and other risks exist for food or energy production and supply. The duality of water (in)security has attracted the attention of different stakeholders, and the idea of being insecure has raised the need to consider risk and consequently the need to measure and quantify water security (ADB, 2016; Varady et al., 2016). The need to quantify, in turn, has led to the creation of numerous qualitative and quantitative indicators. Even so, these metrics may fail to capture the breadth of what it means to be water-secure (Gerlak et al., 2018; Varady et al., 2016).

Not only the use of the term, but research on water security also has increased substantially over the last decade (Bakker, 2012; Gerlak et al., 2018). As climate change and urban infrastructure development occasion more frequent and intense flooding in some regions and more prolonged droughts in others, concerns for water security will almost certainly continue to grow.

From water security to the WEF security nexus: why the nexus is water-centric

Water management has historically had integrationist tendencies. A rudimentary nexus approach to water management can be traced to the 1933 creation of the Tennessee Valley Authority (TVA) in the south-eastern United States. The TVA was mandated to holistically manage water resources, energy generation and agricultural production, while promoting economic development. Its management model was copied and employed for river-basin development in South America, Asia and Africa. Use of this approach gradually declined and was supplanted in the 1990s by the Integrated Water Resources Management (IWRM) framework, which placed greater focus on governance and decision-making. The nexus approach differs from IWRM because it links productive sectors (water, energy, food) and, increasingly, earth systems (de Grenade et al., 2016) to broader human security – a policy integration that highlights resource security. IWRM and nexus modes also differ in scale, participation, and resource use. Nonetheless, water-resource analysts with experience employing integrationist IWRM approaches recognized early on the clear links that water management and policy had with energy and food; this awareness prompted the genesis of the nexus from the water sector (Scott et al., 2015).

In the current context, we contend that at least two features of water – as a resource and domain of management and research – help explain why the water-energy-food nexus construct emerged from the field of water resources and it continues to be viewed from a water-centric perspective. First, water as a ubiquitous resource – even though unevenly distributed across regions and over time – is fundamental to human quality of life and ecosystem processes. Even though there are multiple sources of water (surface water, groundwater, reclaimed water, desalinated water, etc.), they are not substitutable in the same way energy is – consider electricity vs. fuel for heating or transportation, or gas replacing or complementing petroleum in the current fracking boom. And food, too, takes many forms of resource provisioning – crop agriculture, agroforestry, livestock, fisheries and aquaculture, among others. The substitutability and diversification options for water resources remain more limited, in part due to real constraints on long-distance conveyance of water. It is clear that water is a more basic input to energy generation and food production and provisioning than either of the other two resources is for water use and management, even though energy dependence for water security is a rapidly growing concern. Thus, the pursuit of water security is first and foremost a proximate, local challenge, even though the institutions for water management are often established and operate at regional and national scales, with global influences.

Second, the institutions for water governance tend to be more deeply rooted in the public domain, with access to water defined in administrative, regulatory and legal terms. In contrast, energy production and supply – even the pursuit of energy security – are highly commercial endeavours by both private-sector and public-utility interests. This is also increasingly the case for food production, with such features as commodity chains, supply and distribution of inputs, marketing and profits increasingly dominated by mega-scale agri-businesses, many of them global in their scope, reach and impacts. The implication for water resources and water security is that private, commercially oriented actors must turn to public institutions that regulate access to water resources, allocate and oversee water rights, enforce water quality and ecological guidelines, and are the ultimate recourse to disaster relief in the case of drought and flood. Often these functions are beyond the institutional capability of public institutions charged to deliver them; nonetheless, the locus of accountability remains in the public or collective domain. The result is that water security and the public institutions that govern it make water the 'more equal among equals'. Still, it should be noted that this priority-of-place for public governance of water is decreasing as private commercial entities are increasingly involved in water service provision.

Strategies to address water security, whether they aim to develop new water supplies, increase water storage, encourage reduced consumption, or improve water-use efficiency, can have varying implications for energy and food security. As water scarcity and uncertainty of supply increase due to depletion of existing sources, economic growth and climate change, water security will become increasingly critical. Researchers employ water security as a multifaceted goal – most often it includes quantity, quality, and access to water supplies, and addresses the risks involved with overabundance of water, such as flooding (Bakker, 2012; GWP 2000; Grey & Sadoff, 2007), and variability in water supplies due to climate change. To address water security as a broad concept, strategies to achieve water security are multiple and diverse. In some cases, they concurrently promote energy and food security, but not always; there are instances when prioritizing water security can actually detract from these goals. We explore these types of interdependencies in the following section.

Addressing water security: the WEF nexus lens

Water security is often addressed through supply expansion – developing new water supplies or transporting water from water-rich to water-poor areas. Generally, addressing water security through increased supply aids the goal of food security – irrigated agriculture is the largest global water user, and water demands for irrigation are expected to increase (Rosegrant, Ringler, & Zhu, 2009; Simonovic, 2017). Further, alongside global economic development, demand for water-intensive food products, such as cereal grains and meat, is rising rapidly. Water supply, not arable land, is expected to be the key limiting factor in global food security (UNDP, 2007). Increasing water supplies can support energy production by making more water available for electrical plant cooling – water used for thermoelectric cooling made up 45% of all water withdrawals in the US in 2010 (Maupin et al., 2014). However, complex trade-offs between water and energy arise when new water supplies are energy-intensive to develop (e.g., desalination). Further, water scarcity can strain competition for supply among multiple water users.

In many areas experiencing water scarcity, particularly arid and semi-arid environments, easily accessible sources, such as surface water and rainfall, are already fully utilized and sometimes even overallocated. To increase supplies, these areas must develop less accessible sources of water, such as groundwater and desalinated water, or transport water over long distances. However, groundwater extraction requires energy for pumping, as shallow groundwater sources are being used to capacity; expanding groundwater supplies often leads to the need for deeper wells or to developing brackish-water repositories, which require more energy to install and operate. Long-distance water conveyance and desalination are both expensive and energy-intensive (Siddiqi & Anadon, 2011). In Arizona, for example, water from the Colorado River is transported 336 miles in the Central Arizona Project aqueduct – the state's largest single energy user – to provide water to agriculture, cities, mines, and other water users.

Desalination is increasingly used in arid regions to increase water supplies, even though it is expensive, energy-intensive and environmentally damaging, and may generate public opposition based on environmental, cost and equity concerns (Wilder et al., 2016). In the Middle East, Arabian Gulf countries and Israel rely on desalination, wastewater reuse and groundwater to meet water demands (Ghaffour, Missimer, & Amy, 2013; Nair & Kumar, 2013; Tenne, 2013). Technological improvements, reduced costs and increased water demands have made desalination a viable water supply in Gulf countries (Ghaffour et al., 2013). However, 5–12% of total electricity generated in this region is used to power pumps for groundwater extraction and desalination plants (Siddiqi & Anadon, 2011). The quantity of water supplied by desalination in the Middle East and North Africa is expected to nearly double by 2030, expanding demands on energy supply to support water security (Bazilian et al., 2011). Thermal desalination plants are typically coupled with power plants, but whereas water demand is relatively constant throughout the year, lower energy demands in the winter in this hot, arid region make desalination in winter more expensive than in summer (Ghaffour et al., 2013). Storing excess quantities of desalinated water during the summer season, or supplementing with treated wastewater, can help defray extra production costs in winter.

Wastewater treatment is less costly than desalination (Tal, 2016), but in current practice, treated wastewater has limited applications, such as landscape irrigation. Widespread adoption of potable reuse of wastewater is not expected, at least in the coming decades. But expanding the use of treated wastewater to crop irrigation would extend water supplies and address food security. Reuse of 86% of Israel's wastewater provides half of nationwide irrigation water supplies (Tal, 2016). However, the use of treated wastewater for crops presents concerns regarding public health, environmental impacts and social acceptability that may vary regionally and culturally (Qadir, Bahri, Sato, & Al-Karadsheh, 2010).

Improving water-use efficiency and using water supplies of various qualities can help optimize water allocations and free up higher-quality supplies for higher-value or more appropriate uses, such as drinking water. Because irrigated agriculture accounts for 80% of global water use to produce 40% of global food supplies (Molden et al., 2010; Rosegrant et al., 2009), improving irrigation efficiency can significantly expand water supplies for food production and reduce energy requirements for water extraction and conveyance. However, while irrigation-technology improvements can potentially alleviate water stresses in water-scarce regions, some of these technologies, such as drip irrigation, consume more energy than conventional flood irrigation methods (Siddiqi & Anadon, 2011). As supplies of freshwater become increasingly limited, the use of saline water, such as treated wastewater and brackish groundwater, will be increasingly important for use in energy-production processes (Scott & Sugg, 2015). But increasing water supplies through tertiary-level wastewater treatment or desalination has its own trade-offs, due to the energy requirements for collection, treatment and redistribution (Hussey & Pittock, 2012; Yillia, 2016). Further, reducing water needs for energy production is complicated. While new technologies for thermo-electric plants, such as closed-loop systems, withdraw less total water, more water is put to consumptive uses (Sovacool & Sovacool, 2009).

Water scarcity in arid areas also induces planners to seek market-based solutions to expand supplies, for instance, importing water-intensive crops to reduce domestic agricultural water demand. 'Virtual water' gained through importing water-intensive goods can help water-scarce nations reduce local water consumption (Siddiqi & Anadon, 2011), particularly for low-value water uses, such as agriculture (Ghaffour et al., 2013). But, while this market-based solution may reduce national water demands and improve food security, foreign imports to supplement water and food supplies increase international dependence and potentially impose international trade restrictions on these essential supplies (Hanjra & Qureshi, 2010). Furthermore, virtual water is energy-intensive due to the transportation required. As an alternative, viewed from a nexus perspective, local forms of food production, including urban agriculture, can provide food using less energy in transportation. And the cost of virtual-water imports of products may further disadvantage poor communities, who may not have access to these alternatives.

The uneven spatial distribution of precipitation and dramatic seasonal variability of precipitation and streamflow – functions of earth system processes – can cause localized or periodic water scarcity, even in places where total water supply is adequate for annual average demand (Biggs, Duncan, Atkinson, & Dash, 2013; Perrone & Hornberger, 2016). Mismanagement of water supplies – for example, over-extraction and long-distance transfer of groundwater – can make it difficult to meet water demands even where water is not physically scarce. Further, climate change is expected to increase the frequency of precipitation extremes, such as floods and droughts, both of which impact water and food security, particularly in cities. For example, when combined sewer systems overflow (as a consequence of severe storm events), water resources can become polluted. This may require end-of-pipe water treatment plants – which consume additional energy – before runoff reaches rivers and lakes.

In some areas, such as the Himalayan mid-hills, where the number of small hydropower facilities has rapidly grown in mountain basins to capture renewable energy, extreme flooding can disrupt energy production by damaging hydropower facilities through sediment clogging, and destroy irrigation piping for rural villages (Chopra, 2014). Flood protection may also have indirect positive benefits for both energy and food security. Early-warning systems or reinforced infrastructure secure small hydropower plants and farmer-managed irrigation systems in mountainous regions.

Understanding the spatial and temporal relationship between water, food and energy security at the local scale is essential for the security of these resources. For example, in Nepal there are 1977 micro-hydel plants, producing between 10 and 100 kW of power each, distributed up and down river valleys (Surendra, Khanal, Shrestha, & Lamsal, 2011). In these same locations, about 15,000 decentralized, farmer-managed irrigation systems cover about half of the total surface water irrigation in the country (DOI, 2007; Pradhan, 1989). The hydel plants require specific topography and are generally built in hilly or mountainous regions where the river water is shared with other competing uses, mainly irrigation and drinking water. Almost all the hydel plants in Nepal are 'run-of-the-river' systems that have minimal water storage capacity and instead return diverted streamflow back to the river after energy production. As a result, they are highly sensitive to seasonal variation of the river discharge. Since Nepal receives 80% of its annual precipitation during the monsoon season from mid-June to September, there is no competing water demand during that period. But there can be serious competition for the water during the pre-monsoon (early June) and post-monsoon season, when the river discharge is near its lowest and agricultural water demand is at its highest. The location of the hydel plants and irrigation system also plays a key role in determining the interlinkages. Dam structures can disturb the river flow and increase the silt load in the canal infrastructure. All these situations are further amplified by climatic change and variability in the region, which is likely to change the timing and intensity of climatic events.

Temporal variability of precipitation also impacts energy security in economies that rely heavily on water-based energy sources. For example, Brazil is highly dependent on hydropower for electricity supply: hydropower provided over 75% of Brazil's power in 2011 (EIA, 2014). Thus, the nation's electricity supply is vulnerable to changes in precipitation timing and volumes; and precipitation is expected to decrease based on global climate change scenarios (de Lucena et al., 2009). Recurrent droughts in the Amazon River basin, where the majority of Brazil's hydropower is produced, hamper hydropower production to the extent that thermal plants are needed to supplement supply (EIA, 2014). In recent years, reservoirs have been at their lowest levels since 2001 (Prado et al., 2016). Although much of Brazil's hydropower is produced from large-reservoir facilities that have sufficient storage to buffer the effects of short-term drought, increasing environmental regulation has prompted hydropower expansion to shift towards smaller, run-of-the-river designs (de Lucena et al., 2009). Because run-of-the-river hydropower facilities do not impound water, they have less environmental impact (de Lucena et al., 2009). However, because they do not have water storage, they are more vulnerable to drought and shifting rainfall regimes. Brazil is working to expand alternative renewable energy sources, such as biofuels, wind and solar – all with their own trade-offs – to supplement energy supplies from hydropower during droughts and low-flow seasons, to address energy security. Though many renewables are more water-efficient than conventional fossil-fuel-based energy generation, supplementation alone will provide limited solutions as demand continues to expand.

A soft-path approach to water security and WEF trade-offs

Extending water supplies through supply expansion and efficient use addresses the need for adequate water quantity. However, providing access to water sources and preserving sufficient water quality – also critical components of water security – necessitates effective water policy and management. Despite a historic preference for infrastructure solutions such as dams, aqueducts and canals that expand water supplies (Hussey & Pittock, 2012), water security can also be addressed through soft-path approaches such as conservation, increased efficiency, and improved water-use policies (Varady, Zuniga-Teran, Gerlak, Staddon, & Albrecht, in preparation). Water-use efficiency can be raised by increasing water reuse, improving agricultural productivity, rehabilitating infrastructure, and re-allocating supplies to higher-value uses (Rosegrant et al., 2009).

The complex trade-offs among water, energy and food require integrated policy to achieve optimized outcomes in all sectors, not just one. Often, challenges of water policy and governance have more significant impacts on water resource availability than actual resource scarcity. For instance, economic water scarcity can occur due to poor water management. India relies heavily on groundwater for irrigation, extracting some 230 billion m^3 each year to supplement irrigation of 45 million hectares (Shah, Giordano, & Mukherji, 2012). Irrigators throughout western India rely on groundwater, due to its local availability. Initial groundwater declines, due to high withdrawal rates and an expansion of food production, caused drilling and pumping energy demands to increase, due to the greater depths needed to reach groundwater (Scott et al., 2015). To increase water access and improve agricultural productivity, government-supported electricity subsidies were put in place in the mid-1970s. But the energy subsidies caused significant depletion of groundwater aquifers by facilitating overuse. Over-extraction of groundwater results in lower aquifer levels, after which groundwater pumping requires more energy to lift water from greater depths. Irrigation accounts for approximately one-fifth of total electricity use in India. While initially, electricity subsidies improved food security, in the long term, they created a cycle of groundwater depletion and increased energy demand, which nearly bankrupted the power sector in India (Shah et al., 2012).

Not all sources of energy and food supplies compete for water allocations. Rainfed agriculture, including gravity irrigation from rainwater harvesting, makes up the majority of global agriculture practices. Rainwater harvesting has long been used to adapt to drought conditions, providing both household water needs and irrigation water (Ghimire & Johnston, 2015). Also, renewable energy technologies, such as solar photovoltaics and wind power, have relatively low water footprints (Scott et al., 2015). Increased use of solar photovoltaic power could alleviate the energy demands of irrigation and water-supply pumping (Bazilian et al., 2011).

Water security challenges vary geographically, and similarly, trade-offs among water, energy and food security depend on local context. Because challenges vary among locations, strategies to achieve water security draw from a broad range of approaches tailored to local needs and resource availability (Garrick & Hall, 2014; Yillia, 2016). An integrated, nexus approach that considers the availability of water sources and implications for energy and food security will help in developing locally relevant integrated resource policies. Yet, significant barriers to integrated policy exist, such as insufficient information on resource use and trade-offs, existing fragmented sectoral policies, and path dependency in policy-making and governance (Hussey & Pittock, 2012). The ways the goal of water security can limit food and energy security can be minimized through diversification of water, energy and food sources, efficient allocation and expanded institutional capacity. First, diversification of sources provides system resilience, so that, for example, drought years do not cripple both the water and energy sectors. Energy and food supplies should be spread between water-intensive and water-efficient sources. Second, efficient allocation expands water supplies by fulfilling water needs for competing uses with the lowest water quality necessary for that application (also producing energy savings). Third, institutional capacity is critical for examining and maximizing intersectoral trade-offs, using a combination of hard- and soft-path approaches. A nexus approach emphasizes an interdisciplinary perspective, considering not only whether resource quantities fulfil the demand, but also the social, economic, political and environmental implications of resource production and allocation. To minimize intersectoral trade-offs, system optimization requires adequate knowledge of intersectoral interactions, an understanding of cross-sectoral benefits and implications, and consideration of resource use at the planetary level.

From resource security to earth system resilience

As resources, food, energy and water are considered basic human rights. However, earth systems are the fundamental basis for the existence of life on the planet. Earth systems are the series of interlinked, dynamic planetary processes fundamentally interlinked with the WEF nexus, with critical implications for human well-being and ecological resilience, and synergies and trade-offs at all scales. The WEF security nexus characterized in the foregoing discussion often fails to account explicitly for the ecological underpinnings of water, energy and food, and hence, we also consider the concept of earth system resilience.

The conceptual framing of resource security – focusing on food, energy, and subsequently water – emerged under the broader umbrella of environmental security as a foundational component of human security in the late twentieth century (Brown, 1977; Dalby, 2009; Litfin, 1999; Mische, 1989; Myers, 1993; Ullman, 1983). Environmental-security approaches address threats to human well-being and livelihoods from natural hazards and disasters, environmental degradation, limits to natural resources, and global change (Scott & Thapa, 2015; Westing, 2013). Buzan (1991, p. 38) writes insightfully, 'Environmental security concerns the maintenance of the local and the planetary biosphere as the essential support system on which all other human enterprises depend.' Dimitrov (2002) describes human security as inextricable from the health of the ecosystem. The concepts of natural and human-induced global limits to resources are furthered in the planetary boundaries literature, which identifies and measures the risk to earth system function and resilience in the Anthropocene (Steffen et al., 2015). The planetary boundaries approach provided a framework to fundamentally shift from the nebulous, all-encompassing concept of 'the environment' to specific, identifiable earth system processes that can be identified, researched and assessed as underpinning the resources essential to human life and well-being as well as the long-term resilience of all life. Even the planetary boundaries approach is limited, however, in identifying relationships among systems and how changes in one may have multiple unpredictable effects on others.

Addressing earth system resilience in the Anthropocene through research, policy and practice may be the most critical challenge ever faced by humanity. Our human actions seem insignificant in relation to the magnificent, complex and immense planet we inhabit, but we are witnessing the extinction of species that cannot tolerate the novel and toxic ecologies generated by human forces. Biodiversity is the underpinning of a resilient food system, through the broad base of genetic resources and complex agro-ecological systems that sustain our appetites, our bodies and our societies. The carbon cycle is a substantial component of energy security and the fundamental basis of water security. Without geologic, biologic and atmospheric processes of nutrient cycling, soil building, water and air filtering and oceanic circulation, among thousands of fine- to broad-scale systems, life on this planet would not exist, and under the changing systems already active, may be forced to fundamentally evolve or go extinct.

We, as humans, are generating and accentuating dynamic fluctuations in earth system processes, and we also have the ability to imagine, design and implement changes in our behaviour as individuals and societies to moderate the extremes of natural and human-aided system changes. Sustainability as a concept may be applied to resource use and availability as it may be applied to earth systems. But the idea of 'sustaining' falls short of imagining the reaction of systems to the stress and strain of forces caused or accentuated by humans. Along with a new vision of earth systems as a critical fundamental component of nexus approaches, the concept of resource security is shifting to the more agile concept of *resilience* (understood here as social-ecological system reorganization or recovery after shock or disturbance – Gunderson et al., 2010). Resource or even environmental security, especially involving multiple, interlinked resources across multiple sectors in a highly dynamic context, does not account for adaptive action and adaptive capacity in social and ecological systems. The nexus security framework has thus far proved limiting in the analysis of continually changing systems, in which iterative, integrative and participatory approaches are essential to engage shifting and accelerating demands, planetary boundaries and our unlimited imagination in addressing social-ecological challenges. The earth system framework may inherently consider extremes in human and natural systems, including resource availability, future climate scenarios, natural and human-caused disasters, risk, and human conflict.

The subject of earth systems is the emerging critical factor in the nexus discussion; but how earth system resilience (linked to the nexus) is addressed in science and policy has yet to be clearly articulated. Biosphere integrity, biogeochemical flows, land-system change, freshwater use and climate change regulate earth system functions, and may therefore be considered critical factors that sustain all of humanity. These processes do not fit neatly into the triad of water, energy and food security; they underpin and enable all resource security.

While measuring earth system processes remains elusive, the imperative of prioritizing earth system resilience in science and policy is undeniable and urgent. Incorporating earth system resilience as a component of nexus research underscores the need for a conceptual and quantitative understanding of the interplay of multiple resources, institutions and securities. As alluded to above, ecological, hydrological and climate system dynamics as earth system processes are potent drivers of water, energy and food resource use; they also force societal adaptation to global change, and instil further cycles of resource use and dependence.

The appeal of the nexus approach in research, policy and practice is its inherent consideration of multiple, competing and/or synergistic sectors. This approach provides scientists a framework to analyze cross-sectoral interrelationships, and allows policy-makers to focus on integrative policies that increase efficiencies across sectors. Participatory methods of resource planning and management have strengthened the nexus appeal, as NGOs, decision-makers, and stakeholder groups have employed nexus tools to address real-world challenges. While food, water and energy security have been the most common fields of assessment, the nexus framework has also been used to imagine different futures under earth system change, e.g., variable climate and more extreme hydrological processes, and societally induced change, e.g., increased waste production and heightened land-use pressures.

Conclusion: whither the resource securities nexus and earth system resilience?

The resource nexus has been developed as a way to address potential trade-offs in resource demands. Though the nexus concept emerged from a water-sector approach, the nexus considers multiple, synergistic (or competing) cross-sectoral interactions and comprehensive, multi-resource policies. Using this approach, scientists have a framework to assess intersectoral linkages, and decision-makers can focus on integrative policies that better address efficiencies across sectors.

Security as a nexus concept has an implied achievable end goal, whereas resilience invites a focus on adaptive action and adaptive capacity in social and ecological systems. With the resilience dimension, nexus resource availability, environmental hazards, and the overall function and services of interlinked earth systems must be balanced. In this sense, nexus approaches should be process-oriented and address dynamic interplays in research and policy. These approaches should also be flexible enough to account for extreme environmental and social conditions, which will become more commonplace with global change.

The challenge lies in strengthening nexus-security-resilience linkages and giving these greater applicability to enable more meaningful operational results. To do so, we must first consider how security discourses have been framed and to what effect. Water security has provided an integrative approach through which the resource nexus has emerged in science and policy. Yet, strengthening nexus-security-resilience linkages will be crucial in the coming decades as threats of food security, energy security and water security heighten in many parts of the world due to anthropogenic and hydroclimatic processes, which clearly indicate the need to more holistically consider earth system processes.

Disclosure statement

No potential conflict of interest was reported by the authors.

Funding

This work was undertaken with partial support of the International Water Security Network, a project funded by Lloyd's Register Foundation, a charitable foundation helping protect life and property by supporting engineering-related education, public engagement and the application of research. Additional support was provided by the Inter-American Institute for Global Change Research (project CRN3056), which is supported by the US National Science Foundation (grant no. GEO-1128040), NSF grant no. DEB-1010495, and the Morris K. Udall and Stewart L. Udall Foundation in Tucson, Arizona.

References

Albrecht, T. R., Crootof, A., & Scott, C. A. (2018). The water-energy-food nexus: A systematic review of methods for nexus assessment. *Environmental Research Letters, 13*, 043002.

Asian Development Bank. (2016). *Asian water development outlook 2016: Strengthening water security in Asia and the Pacific*. Manila, Philippines: Asian Development Bank.

Bakker, K. (2012). Water security: Research challenges and opportunities. *Science, 337*, 914–915.

Bazilian, M., Rogner, H., Howells, M., Hermann, S., Arent, D., Gielen, D., ... Yumkella, K. K. (2011). Considering the energy, water and food nexus: Towards an integrated modelling approach. *Energy Policy, 39*(12), 7896–7906.

Benson, D., Gain, A., & Rouillard, J. (2015). Water governance in a comparative perspective: From IWRM to A 'nexus' approach? *Water Alternatives, 8*(1), 756-773.

Biggs, E. M., Duncan, J. M. A., Atkinson, P. M., & Dash, J. (2013). Plenty of water, not enough strategy. How inadequate accessibility, poor governance and a volatile government can tip the balance against ensuring water security: The case of Nepal. *Environmental Science and Policy, 33*, 388–394.

Brown, L. (1977). Redefining national security. *Worldwatch Paper 14*, p. 46.

Buzan, B. (1991). *People, states, and fear: An agenda for international security studies in the post-Cold War era*. Boulder, CO: Lynne Rienner.

Cherp, A., & Jewell, J. (2011). The three perspectives on energy security: Intellectual history, disciplinary roots and the potential for integration. *Current Opinion in Environmental Sustainability, 3*(4), 202–212.

Chopra, R. (2014). *Uttarakhand: Development and ecological sustainability* (Report produced for Oxfam India). New Delhi: Oxfam India.

Cook, C., & Bakker, K. (2012). Water security: Debating an emerging paradigm. *Global Environmental Change, 22*(1), 94–102.

Dalby, S. (2009). *Security and environmental change*. Cambridge: Polity.

de Grenade, R., House-Peters, L., Scott, C. A., Thapa, B., Mills-Novoa, M., Gerlak, A., & Verbist, K. (2016). The nexus: Reconsidering environmental security and adaptive capacity. *Current Opinion in Environmental Sustainability, 21*, 15–21.

de Lucena, A. F. P., Szklo, A. S., Schaeffer, R., de Souza, R. R., Borba, B. S. M. C., da Costa, I. V. L., ... da Cunha, S. H. F. (2009). The vulnerability of renewable energy to climate change in brazil. *Energy Policy, 37*(3), 879–889.

Dimitrov, R. S. (2002). Water, conflict, and security: A conceptual minefield. *Society and Natural Resources, 15*(8), 677–691.

DOI. (2007). *Development of database for irrigation development in Nepal* (pp. 180). Kathmandu, Nepal: Department of Irrigation.

EIA. (2014). Hydropower supplies more than three-quarters of Brazil's electric power. U.S. Energy Information Administration. Retrieved from https://www.eia.gov/todayinenergy/detail.php?id=16731

Garrick, D., & Hall, J. W. (2014). Water security and society: Risks, metrics, and pathways. *Annual Review of Environment and Resources, 39*, 611–639.

Gerlak, A. K., House-Peters, L., Varady, R., Albrecht, T., Zuniga Teran, A., de Grenade, R., … Scott, C. A. (2018). Water security: A review of place-based research. *Environmental Science and Policy, 82*, 79–89.

Ghaffour, N., Missimer, T. M., & Amy, G. L. (2013). Combined desalination, water reuse, and aquifer storage and recovery to meet water supply demands in the GCC/MENA region. *Desalination and Water Treatment, 51*, 38–43.

Ghimire, S. R., & Johnston, J. M. (2015). Traditional knowledge of rainwater harvesting compared to five modern case studies. *World Environmental and Water Resources Congress 2015: Floods, Droughts, and Ecosystems*, 182–193. doi:10.1061/9780784479162.017

Global Water Partnership. (2000). *Towards water security: A framework for action*. Stockholm: Global Water Partnership.

Godfray, H. C. J., & Robinson, S. (2015). Contrasting approaches to projecting long-run global food security. *Oxford Review of Economic Policy, 31*(1), 26–44.

Grey, D., & Sadoff, C. W. (2007). Sink or swim? Water security for growth and development. *Water Policy, 9*(6), 545–571.

Gunderson, L. H., Allen, C. R., & Holling, C. S. (Eds.). (2010). *Foundations of ecological resilience*. Washington, DC: Island Press.

Hanjra, M. A., & Qureshi, M. E. (2010). Global water crisis and future food security in an era of climate change. *Food Policy, 35*(5), 365–377.

Hoff, H. (2011). *Understanding the nexus, background paper for the Bonn 2011 conference: The water, energy and food security nexus*. Stockholm: Stockholm Environment Institute.

Hussey, K., & Pittock, J. (2012). The energy-water nexus: Managing the links between energy and water for a sustainable future. *Ecology and Society, 17*(1). doi:10.5751/ES-04641-170131

Ingram, J., Ericksen, P., & Liverman, D. (Eds.). (2010). *Food security and global environmental change*. London: Earthscan.

Kumar, M. D., Bassi, N., Narayanamoorthy, A., & Sivamohan, M. K. V. (Eds.). (2014). *The water, energy and food security nexus*. Abingdon: Routledge.

Kuyper, T. W., & Struik, P. C. (2014). Epilogue: Global food security, rhetoric, and the sustainable intensification debate. *Current Opinion in Environmental Sustainability, 8*, 71–79.

Litfin, K. T. (1999). Constructing environmental security and ecological interdependence. *Global Governance, 1999*, 359–377.

Maupin, M. A., Kenny, J. F., Hutson, S. S., Lovelace, J. K., Barber, N. L., & Linsey, K. S. (2014). Estimated use of water in the United States in 2010: U.S. Geological Survey Circular 1405. p.56. Retrieved from https://pubs.usgs.gov/circ/1405/

McGlade, C., & Ekins, P. (2015). The geographical distribution of fossil fuels unused when limiting global warming to 2°C. *Nature, 517*, 187–190.

Meadows, D. H., Meadows, D. L., Randers, J., & Behrens, I. I. (1972). *The limits to growth*. New York, NY: Universe Book.

Mische, P. M. (1989). Ecological security and the need to reconceptualize sovereignty. *Alternatives XIV, 14*(4), 389–427.

Molden, D., Oweis, T., Steduto, P., Bindraban, P., Hanjra, M. A., & Kijne, J. (2010). Improving agricultural water productivity: Between optimism and caution. *Agricultural Water Management, 97*(4), 528–535.

Myers, N. (1993). *Ultimate security: The environmental basis of political stability*. New York, NY: W. W. Norton.

Nair, M., & Kumar, D. (2013). Water desalination and challenges: The Middle East perspective: A review. *Desalination and Water Treatment, 51*(10–12), 2030–2040.

Nelson, M. C., Ingram, S. E., Dugmore, A. J., Streeter, R., Peeples, M. A., McGovern, T. H., … Smiarowski, K. (2016). Climate challenges, vulnerabilities, and food security. *Proceedings of the National Academy of Sciences, 113*(2), 298–303.

O'Rourke, D., & Connolly, S. (2003). Just oil? The distribution of environmental and social impacts of oil production and consumption. *Annual Review of Environment and Resources, 28* (1), 587–617.

Pachauri, S., & Cherp, A. (2011). Energy security and energy access: Distinct and interconnected challenges. *Current Opinion in Environmental Sustainability, 3*(4), 199–201.

Pasqualetti, M. J., & Sovacool, B. K. (2012). The importance of scale to energy security. *Journal of Integrative Environmental Sciences, 9*(3), 167–180.

Perrone, D., & Hornberger, G. (2016). Frontiers of the food–energy–water trilemma: Sri Lanka as a microcosm of tradeoffs. *Environmental Research Letters, 11*(1), 1–10.

Pradhan, P. (1989). *Patterns of irrigation organization in Nepal: A comparative study of 21 farmer-managed irrigation systems* (No. 1). Colombo, Sri Lanka: International Water Management Institute.

Prado, F. A., Athayde, S., Mossa, J., Bohlman, S., Leite, F., & Oliver-Smith, A. (2016). How much is enough? An integrated examination of energy security, economic growth and climate change related to hydropower expansion in Brazil. *Renewable and Sustainable Energy Reviews, 53*, 1132–1136.

Qadir, M., Bahri, A., Sato, T., & Al-Karadsheh, E. (2010). Wastewater production, treatment, and irrigation in Middle East and North Africa. *Irrigation and Drainage Systems, 24*(1–2), 37–51.

Rosegrant, M. W., Ringler, C., & Zhu, T. (2009). Water for agriculture: Maintaining food security under growing scarcity. *Annual Review of Environment and Resources, 34*(1), 205–222.

Sachs, J. D. (2012). From millennium development goals to sustainable development goals. *The Lancet, 379*(9832), 2206–2211.

Salvatici, S. (2015). Writing the history of food security since 1945. *Contemporanea, 18*(3), 347–354.

Scott, C., & Thapa, B. (2015). Environmental security. Retreived from http://www.oxfordbibliographies.com/display/id/obo-9780199363445-0012

Scott, C. A., Kurian, M., & Wescoat, J. L., Jr. (2015). The Water-energy-food nexus: Adaptive capacity to complex global challenges. In M. Kurian & R. Ardakanian (Eds.), *Governing the nexus: Water, soil and waste resources considering global change* (pp. 15–38). Berlin: Springer.

Scott, C. A., & Sugg, Z. P. (2015). Global energy development and climate-induced water scarcity-Physical limits, sectoral constraints, and policy imperatives. *Energies, 8*(8), 8211–8225.

Shah, T., Giordano, M., & Mukherji, A. (2012). Political economy of the energy-groundwater nexus in India: Exploring issues and assessing policy options. *Hydrogeology Journal, 20*(5), 995–1006.

Shaw, J. (2007). *World food security: A history since 1945*. Basingstoke: Palgrave Macmillan.

Siddiqi, A., & Anadon, L. D. (2011). The water-energy nexus in Middle East and North Africa. *Energy Policy, 39*(8), 4529–4540.

Simonovic, S. P. (2017). Bringing future climatic change into water resources management practice today. *Water Resources Management, 31*, 2933.

Sovacool, B. K. (2007). Solving the oil independence problem: Is it possible? *Energy Policy, 35*(11), 5505–5514.

Sovacool, B. K., & Brown, M. A. (2010). Competing dimensions of energy security: An international perspective. *Annual Review of Environment and Resources, 35*(1), 77–108.

Sovacool, B. K, & Sovacool, K. E. (2009). Identifying future electricity-water tradeoffs in the united states. *Energy Policy, 37*(7), 2763–2773. doi:10.1016/j.enpol.2009.03.012

Steffen, W., Richardson, K., Rockstrom, J., Cornell, S. E., Fetzer, I., Bennett, E. M., … Sorlin, S. (2015). Planetary boundaries: Guiding human development on a changing planet. *Science, 347*(6223), 1–10.

Strong, T. H. (1956). Land tenure in australia in relation to technical advances and closer settlement. *Journal Of Farm Economics, 38*(2), 458–464. doi:10.2307/1234385

Surendra, K. C., Khanal, S. K., Shrestha, P., & Lamsal, B. (2011). Current status of renewable energy in Nepal: Opportunities and challenges. *Renewable and Sustainable Energy Reviews, 15*(8), 4107–4117.

Sustainable Development Goals. (2016). GlobalGiving. Retrieved from https://www.globalgiving.org/sdg/

Tal, A. (2016). Rethinking the sustainability of Israel's irrigation practices in the Drylands. *Water Research, 90*, 387–394.

Tenne, A. (2013). Sea water desalination in Israel: Planning, coping with difficulties, and economic aspects of long-term risks. In S. B. Megdal, R. G. Varady, & S. Eden (Eds.), *Shared borders, shared waters: Israeli-Palestinian and Colorado river basin water challenges* (pp. 263–273). Leiden: CRC Press/Balkema – Taylor & Francis.

Theisen, O. M. (2008). Blood and soil? Resource scarcity and internal armed conflict revisited. *Journal of Peace Research*, *45*(6), 801–818.

Ullman, R. H. (1983). Redefining security. *International Security*, *8*(1), 129–153.

UN Water. (2015). Water for food. Retrieved October 13, 2017, from http://www.unwater.org/fileadmin/user_upload/unwater_new/docs/water_for_food.pdf

UNDP. (2007). *Human development report 2006 – Beyond scarcity: Power, poverty and the global water crisis*. New York, NY: United Nations Development Programme.

Varady, R. G., Zuniga-Teran, A. A., Garfin, G. M., Martin, F., & Vicuña, S. (2016). Adaptive management and water security in a global context: Definitions, concepts, and examples. *Current Opinion in Environmental Sustainability*, *21*, 70–77.

Varady, R. G., Albrecht, T. R., Gerlak, A. K., Zuniga, A. A., & Staddon, C. (In preparation). The water security discourse and its main actors. In J. J. Bogardi, K. D. Wasantha Nandalal, R. R. P. van Nooyen, & A. Bhaduri (Eds.), *Handbook of water resources management*. Springer.

Westengen, O. T., & Banik, D. (2016). The state of food security: From availability, access and rights to food systems approaches. *Forum for Development Studies*, *43*(1), 113–134.

Westing, A. H. (2013). *From environmental to comprehensive security*. London: Springer.

Wilder, M. O., Aguilar-Barajas, I., Pineda-Pablos, N., Varady, R. G., Megdal, S. B., McEvoy, J., … Scott, C. A. (2016). Desalination and water security in the US–Mexico border region: Assessing the social, environmental and political impacts. *Water International*, *41*(5), 756–775.

World Food Summit. (1996). Rome declaration on world food security and world food summit plan of action. Retrieved from http://www.fao.org/docrep/003/w3613e/w3613e00.htm

Yillia, P. (2016). Water-Energy-Food nexus: Framing the opportunities, challenges and synergies for implementing the SDGs. *Österreichische Wasser- Und Abfallwirtschaft*, *1–2*, 1–13.

Zeitoun, M., Lankford, B., Krueger, T., Forsyth, T., Carter, R., Hoekstra, A. Y., … Matthews, N. (2016). Reductionist and integrative research approaches to complex water security policy challenges. *Global Environmental Change*, *39*, 143–154.

Unraveling transboundary water security in the arid Americas

Tamee R. Albrecht, Robert G. Varady, Adriana A. Zuniga-Teran, Andrea K. Gerlak, Rafael Routson De Grenade, América Lutz-Ley, Facundo Martín, Sharon B. Megdal, Francisco Meza, Diego Ocampo Melgar, Nicolás Pineda, Facundo Rojas, Rossi Taboada and Bram Willems

ABSTRACT

Transboundary waters are characterized by diverse and complex socio-politico-economic obstacles to effective water management. We examine five distinct cases in the arid Americas – in locations from the US–Mexico border to the Andes mountains – employing water security as a conceptual prism to unravel the multiple and varied attributes of transboundary water challenges. We describe how borders complicate water security in arid regions and explore how institutional arrangements and practices – within and across jurisdictions – respond to these challenges. We find that institutional capacity is needed on multiple levels for effective water management, and institutions must be responsive and flexible to change.

Introduction

Physiographic regions – defined by unique geography, climate and ecosystems – often span human-drawn borders of nation-states or other political jurisdictions across which governance structures and processes may diverge (e.g., state, municipal, tribal). While such transboundary regions are divided politically and administratively, their ecosystems and natural resources – and water resources in particular – remain connected.

Until the 1980s, discussion of issues peculiar to these cross-border regions usually focused on surface water in the context of large international catchments, with

Supplementary data for this article can be accessed at https://doi.org/10.1080/02508060.2018.1541583

disagreements over allocation of their waters spawning river-basin commissions and sometimes treaties. But other than in regions of potential conflict, the literature on natural-resources management infrequently evoked transboundary issues.[1] Instead, management at a country or jurisdiction level was emphasized. As scholars and water professionals began to appreciate the prevalence and significance of such basins (Udall & Varady, 1994; Varady & Morehouse, 2003; Wolf, Natharius, Danielson, Ward, & Pender, 1999), they also recognized that transboundary situations are characterized by socio-politico-economic obstacles – e.g., asymmetrical infrastructure, disparate national values and interests, contrasting institutional arrangements, and the conflicting needs of upstream versus downstream riparian states – that challenge effective management (Kliot, Shmueli, & Shamir, 2001; Marty, 2001; Morehouse, 1995).

Over the past three decades, approaches to transboundary water management have emphasized nation-to-nation cooperation, international treaties and agreements, and multinational organizations. All these instruments have been proposed to address challenges of hegemonic control, equitable use, conflict resolution, and coordination – with transboundary rivers the most common objects of these efforts. Early studies include Wolf (1998), Hamner and Wolf (1998), Falkenmark and Lundqvist (1999), Milich and Varady (1999), Marty (2001), Sadoff and Grey (2002), and Uitto and Duda (2002).[2] In the case of water, these approaches promote local-level participation and individualized treaties to address the site-specific nature of transboundary challenges (Blatter & Ingram, 2001; Conca, 2005; Feitelson & Haddad, 2001; Gerlak, 2004; Ingram, Laney, & Gilliland, 1995; Schmeier, 2013; Van der Zaag & Savenije, 2000; Zeitoun & Mirumachi, 2008; Zeitoun & Warner, 2006). Since the early 2000s, other types of transboundary contexts for water have been increasingly recognized that do not necessarily follow basin boundaries, such as international river-water transfers, effluent exchanges, and import/export of 'virtual water' via water-intensive goods and produce (Allan, 2002, 2011). In short, cross-border water contexts are richly varied and complex.

In this paper, we employ water security as a conceptual prism to unravel the multiple and varied attributes of transboundary water challenges and distinguish the most pertinent water-security attributes for five cases in the arid Americas. Water security is a broad, integrative concept, incorporating many attributes, including aspects of water supply, such as water quantity and quality; water access, affordability and equity; water use for the environment and safeguarding ecosystems; and protection from water-related hazards, such as floods and droughts (e.g., Bakker, 2012; FAO, 1996; GWP, 2000; Grey & Sadoff, 2007; Norman, Bakker, Cook, Dunn, & Allen, 2010; OECD, 2013; Scott et al., 2013; UNESCO, 2013; UN-Water, 2013). A water-security framing facilitates examination of water challenges at multiple spatial scales, and invokes a range of types of institutional capacity to address water challenges at multiple governance levels (Bakker & Morinville, 2013).

Our five cases – in locations from the US–Mexico border to the Andes mountains – feature different geophysical and socio-political transboundary contexts that are thematically distinct. Yet, all are in arid-to-semiarid regions, where local surface water scarcity is a common challenge. Given the complexity introduced when political borders transect resource systems, in addition to the highly place-specific nature of water management, we expect to find a range of multidimensional water-security challenges and an equally broad array of institutional responses.

We begin by describing our case-study approach and the water-security framing technique we employ for analysis. Then we present five distinctive case studies. In addition to some brief background and historical context, we outline the water-security attributes and highlight the key institutional and policy dimensions. Further detail for each case study appears in the online supplementary material (https://doi.org/10.1080/02508060.2018.1541583). In the discussion, we synthesize across the cases with an eye to the commonalities of our case-study research. We conclude with lessons for water-security research in water-scarce transboundary regions of the American continent, while suggesting that those lessons also may be relevant for other arid-to-semiarid regions.

Our approach

Seeing transboundary water issues through a water-security prism

In recent literature, water security has been defined in various ways that incorporate multiple attributes. Conceptual and empirical research emphasize aspects of adequate quantity, sufficient quality, access, ecosystem health, and protection from risk or hazards (Gerlak et al., 2018). Table 1 shows the range of water-security attributes identified in frequently cited definitions appearing in academic literature and policy documents. We adopt a broad approach to water security, emphasizing in-depth, qualitative analysis as opposed to a reductionist, quantitative approach (Zeitoun et al., 2016).

Because of the overriding importance of place, we can expect certain attributes of water security to be more, or less, predominant in different locations, scales and

Table 1. Water security attributes and source definitions (adapted and revised from Gerlak et al. 2018). In each category, attributes are listed in order of frequency. Each source definition included at least six attributes.

	Physical	Quantity Quality Ecosystems Watersheds Global change	
Water Security Attributes	Human	Access Health/Well-being Economic growth Political stability* Policy Livelihoods Sanitation Time/Reliability Preference Transport Industrial resources	*Sources: FAO 1996; GWP 2000; Grey & Sadoff 2007; UNEP 2009; Norman et al. 2010; Zeitoun 2011; Lautze & Manthrithilake 2012; Bakker 2012; OECD 2013; UNESCO 2013; UN-Water 2013; Scott et al. 2013; Jepson 2014; Wheater & Gober 2015; Gain et al. 2016*
	Cross-cutting	Sustainability Risk/Hazards Resilience Uncertainty Food and energy resources	

*Includes peaceful international relationships, intra-national political stability, and national security.

contexts (Cook & Bakker, 2012; Gerlak et al., 2018; Jepson, 2014). Attributes of water security in Table 1 can be argued to be more fundamental (e.g., quantity, quality) or more influential – meaning they have a strong effect on fundamental aspects – for example, resilience to global change (Scott et al., 2013), political stability (UN-Water, 2013), or food and energy policies (Varis, Keskinen, & Kummu, 2017). Many factors can influence water security – and its obverse, insecurity – including culture, politics, power relationships, level of cooperation, infrastructure, and management regimes (Gerlak et al., 2018). In international contexts, peaceful inter-state relations and intra-state stability are particularly relevant to water-security frames (Jepson, 2014). However, transboundary water security involves more than formalized international cooperation, treaties and agreements – it encompasses many aspects of how societies use water resources (Mirumachi, 2013).

To portray the varied attributes of water security demonstrated in Table 1, we conceptualize water security as a prism through which transboundary water issues can be examined. When viewed through the prism of water security, multiple attributes are revealed for each case study – much in the way a dispersion prism separates white light into its constituent colours (Figure 1). We use the water-security prism as an analytical tool, combined with in-depth case study analysis, to distinguish the most pertinent aspects of transboundary water issues for each case from among the many that can characterize water security. The resultant key water-security attributes for each case reflect the highly contextualized and multidimensional nature of water-resource challenges. For example, in some transboundary contexts, the pressures of economic development and political contention may be most critical to understanding water security; in other places, ensuring equitable water access and sanitation may be paramount.

Guided by the attributes of water security, we scrutinize the transboundary character of each case – that is, how a water source (stream, aquifer, glacier) or activity (water transfer) interacts with boundaries to influence water security. Recognizing that the transboundary character of shared water sources is profoundly influenced by how political entities govern and use these resources (Sanchez & Eckstein, 2017), we pay special attention to the geopolitical, institutional and economic aspects of transboundary water utilization – particularly how these conditions can deteriorate and change in response to water-security challenges.

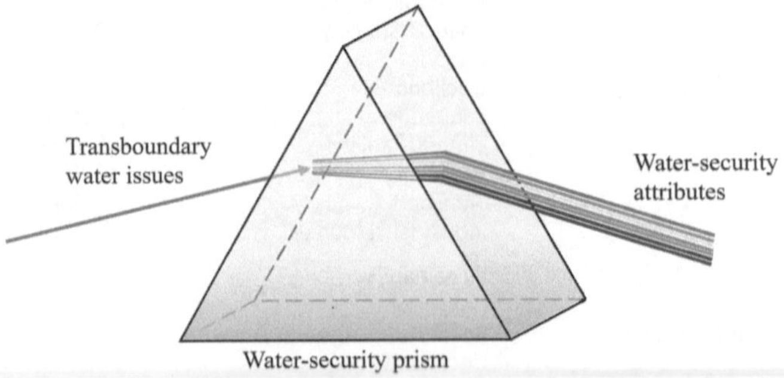

Figure 1. Conceptual diagram of the water-security prism used to separate transboundary water issues into various water-security attributes.

Case-study approach

Following Yin (2003, 2017), we use a qualitative, descriptive case-study approach to determine how water-security challenges are manifested in five illustrative border regions of the arid Americas. We ask, how do borders influence water-security challenges in the arid Americas? And what institutional responses to such challenges can we observe? To address these questions, we identify key water-security attributes that are evident in: (a) water-security challenges, (b) the influence of the transboundary context, and (c) institutional capacity and transboundary cooperation. Knowledge and information on the cases is derived from multiple sources, including local expertise, fieldwork, and attendance at stakeholder workshops, formal meetings, and science-policy dialogues – buttressed by document analysis and literature reviews.[3]

In a north-to-south continuum across the Americas our cases feature an assortment of physical and socio-economic landscapes (Figure 2). Table 2 describes each case's defining features, which include various types of water supplies shared across borders – surface water, groundwater, desalinated water, and glacial sources. They also represent a host of water-security concerns intersecting with instances of cross-border political friction, many of which underscore long-standing geopolitical conflict or disparate ecopolitical power among water users. The cases are all located in semiarid-to-arid environments,[4] where water-resource challenges are pronounced due to surface water scarcity – a condition that is expected to worsen given climate-change projections (IPCC 2007, 2014; Maestre, Salguero-Gomez, & Quero, 2012). In the discussion that follows the case studies, we synthesize lessons learned across the arid-Americas studies.

Five transboundary cases, five water-security challenges

In this section, we discuss each case study, guided by our framing of water security as a spectrum of various attributes (Table 1; Figure 1). For each case, we link our observations to water security by identifying each case's relevant attributes, which are emphasized in italics.

Case study 1. Water security and groundwater in the binational Santa Cruz River basin: cooperation in the contentious US–Mexico border region

Spanning the US–Mexico border, the Santa Cruz River basin encompasses the cross-border Ambos Nogales urban corridor – the neighbouring cities of Nogales, Sonora, and Nogales, Arizona (Figure 3). Here, and throughout the border region, rapid industrialization, urbanization, and population growth have occurred over the past few decades on the Mexican side, fueled by US consumer demands and a comparatively weak Mexican economy – causing uneven development patterns and stressing the natural environment (Varady & Ward, 2009). Economic asymmetries also contributed to periods of high migration rates and a resultant long-standing US security presence at the border (Varady & White, 1992). Despite periodic disagreements between the two countries on certain border issues[5] and stark differences in legal-political systems, when it comes to water, they have historically found ways to cooperate. Viewed through the water-security prism, this case highlights the key attributes of water *quality*, particularly

Figure 2. The American continent, showing the five case studies. Basemap sources: Esri, DeLorme, HERE, MapmyIndia.

as it relates to *sanitation*, as well as water resource *uncertainty* in the face of *global change*. It also brings to the fore how asymmetric *economic growth* and *political* contention complicate transboundary water security.

Water-security challenges

As a semiarid basin with limited surface water, the region relies heavily on groundwater supplies for the cross-border urban corridor, as well as for agricultural water uses on the US side and industrial uses on the Mexican side (Scott, Megdal, Oroz, Callegary, & Vandervoet, 2012a). Easily accessible, shallow groundwater is vulnerable to drought (Shamir et al., 2015), and while deeper stores are sheltered from climatic fluctuations, they are costlier to tap, and their long-term sustainability is largely unknown (Nelson, 2007). On the Mexican side of the border, local groundwater is insufficient to meet

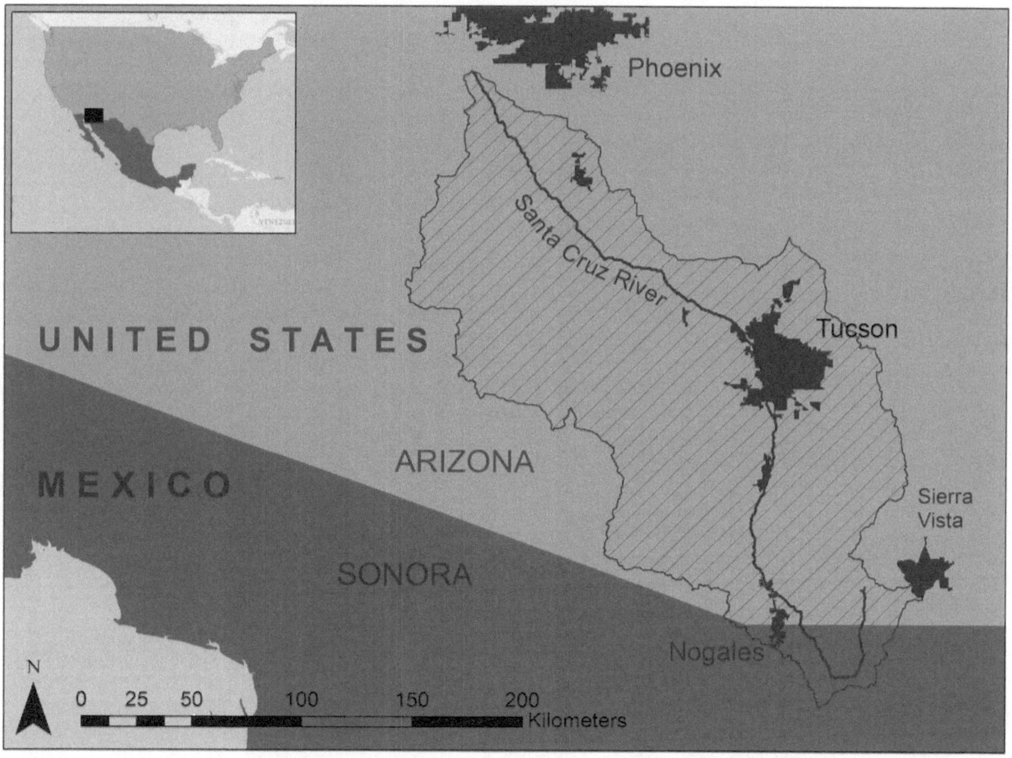

Figure 3. Location of Santa Cruz River basin, crossing the border between Arizona (US) and Sonora (Mexico). The aquifer lies within the basin. Basemap sources: Esri, DeLorme, HERE, MapmyIndia.

demands – up to 40% of the water used in Nogales, Sonora, is transferred from the Los Alisos basin to the south (Wilder et al., 2012).

On both sides of the border, pollution threatens water security. Nogales, Sonora, has undergone rapid economic and population growth over the past two decades due to expansion of the *maquiladora* industry (foreign-owned plants) that boomed in northern Mexico following the passage of NAFTA[6] (EPA & SEMARNAT, 2013). The population in Nogales, Sonora, has grown to over 200,000 (INEGI, 2010) – nearly 10 times the population of its Arizona sister city (US Census, 2015). This rapid growth stresses the natural environment, particularly because wastewater infrastructure development has not kept pace with increases in industrial and domestic wastewater production. Statewide, wastewater-treatment coverage in Sonora is less than 40% (CONAGUA, 2016). In Nogales, large populations live in off-grid residential areas (*colonias*), which typically lack municipal services (Kelly-Richards & Banister, 2017). Leaks from old or overworked pipes, open wastewater discharges in areas where wastewater infrastructure is unavailable, and system overflows that occur during storm events contaminate shallow groundwater and surface water (Kapoor, 2017). Because the growing urban areas of Nogales, Sonora, are upstream of Arizona, wastewater treatment is a transboundary concern.

Table 2. The five case studies at a glance.

Site	Location	Defining issue	Climate	Authors
Upper Santa Cruz Basin	Arizona–Sonora (US–Mexico) border region	Transboundary groundwater	Semiarid Sonoran/Chihuahuan Desert; 35–90 cm annual precip.; most occurs during monsoon (July–September)	Albrecht, Megdal, Varady, Lutz-Ley
Northwest Mexico Desalination	Coastal areas of Sonora and Baja California	Supply enhancement via binational desalination	Semiarid Sonoran Desert and Upper Gulf of Mexico; less than 30 cm annual precip.	Zuniga-Teran, Lutz-Ley, Pineda
Catamayo-Chira River Basin	Western transboundary Ecuador–Peru area	Transboundary river basin spanning a historically contentious international border	High variability in precip. across basin (sea level to 3700 m elevation)	Taboada, Willems
Ica River Basin	South-central Peru	Trans-jurisdictional hydro-hegemony within a single nation	Varies throughout basin from cold and humid in headwaters (4600 m elevation) to dry, desert along Pacific coast; and from 1 cm (lower basin) to 77 cm (upper basin) annual precip.	De Grenade, Taboada
Maipo River Basin (Chile) and Mendoza River Basin (Argentina)	Interglacier river basins in central Chile and Argentina	Montane glacier-fed basins from a shared source-region	Maipo: 29 cm annual precip.; Mendoza: 23 cm annual precip. Both: low-lying areas influenced by strong intra-annual and interannual variability (e.g., ENSO)[a]	Ocampo Melgar, Meza, Martín, Rojas

[a]ENSO: The El Niño Southern Oscillation produces wet El Niño years, in contrast to dry La Niña years (Aceituno, 1989).

Influence of the transboundary context

Groundwater supplies are vulnerable to pollution and overuse on both sides of the border. While researchers suggest that groundwater extraction and pollution in Mexico will affect groundwater availability and quality downstream in Arizona, the extent of such impacts is uncertain (Milman & Scott, 2010). Limited scientific data and an incomplete understanding of the hydrogeology of the Santa Cruz Aquifer, particularly its degree of interconnection across the international border, complicate the task of addressing water security in this transboundary region.

Asymmetries in water governance between the US and Mexico also complicate international water cooperation (Table 3). Mexican water management is highly centralized at the federal level (Scott & Banister, 2008). Water-rights allocation is managed by the Mexico City–based National Water Commission (CONAGUA), which is far removed from transboundary water issues in the northern state of Sonora.

In the US, in contrast, water is mostly governed at the state level, with management tasks often split among multiple agencies (Table 3; Milman & Scott, 2010). For example, in Arizona water quality is in the domain of the Arizona Department of Environmental Quality, while water rights and allocations are administered by the Arizona Department of Water Resources. Management of surface water and groundwater resources is also conducted separately (Megdal & Scott, 2011). Arizona regulates groundwater use via comprehensive legislation designed to maintain 'safe yield'[7] and prevent long-term groundwater-level declines (1980 Groundwater Management Act, Arizona Revised Statutes §45–561). The area of the Santa Cruz basin in Arizona was designated a separate Active Management Area in 1994.

Table 3. Organizations involved in Arizona and Sonora water governance (adapted from Milman & Scott, 2010).

Organization	Role
Arizona	
City of Nogales	Local water utility
Santa Cruz Active Management Area	Regional groundwater planning and management
Water Resources Research Center	State, university-based research institute
Arizona Dept. of Water Resources	Administers and enforces Groundwater Management Act
Arizona Dept. of Environmental Quality	Establishes and enforces water-quality standards
US Geological Survey	National-level scientific research agency
Sonora	
OOMAPAS	Municipal water utility and sanitation provider
Sonora State Water Commission (CEA)	Coordinates water-supply programmes and planning
Organismo de Cuenca Regi6n II Noroeste	Develops regional water plans with public and private users
National Water Commission (CONAGUA)	Governs water rights, allocation and regulation
International	
Transboundary Aquifer Assessment Program	Binational scientific assessment of priority transboundary aquifers
Border Environmental Cooperation Commission	Approves environmental infrastructure and gives technical assistance in border regions
AZ-Mexico Commission and Commission Sonora-Arizona	Non-profit organizations, facilitate trade and public/private collaborations in border region
International Boundary and Water Commission	Interprets and implements the 1944 Water Treaty; provides avenue for binational cooperation, dispute resolution

Institutional capacity and transboundary cooperation

Despite asymmetries, the US and Mexico have a long history of cooperation on water. In 1906, the Boundary Waters Convention was created to resolve transboundary surface water issues on the Rio Grande River. The 1944 Water Treaty addressed international allocations of the Colorado, Rio Grande and Tijuana Rivers, and transboundary sanitation issues.[8] This agreement also established the International Boundary and Water Commission (IBWC),[9] which provides binational institutional capacity, offering a formal pathway for joint processes, dispute resolution, and a mechanism for binational cooperation (Mumme, 2015).

In the Upper Santa Cruz basin, the two countries have cooperated on a binational solution for insufficient wastewater-treatment capacity in Nogales, Sonora – a problem that threatens groundwater on both sides of the border, along with ecosystems and streams. The Nogales International Wastewater Treatment Plant, constructed in 1951 and enlarged, relocated, and modernized several times since, has treated a significant portion of wastewater effluent from Nogales, Sonora, since its inception (Ingram et al., 1995). However, there is still tension between the two countries regarding payment for wastewater treatment services as a consequence of elevated flows sent by Mexico during storm events, and disagreement among US agencies regarding who is responsible for repairs of the leaking conveyance system (Kapoor, 2017).

While there is a tradition of formal binational cooperation on water, existing binational accords do not specifically address shared groundwater.[10] Without a formal agreement on groundwater, cooperation has progressed in other ways. The most meaningful advance is the Transboundary Aquifer Assessment Program (TAAP), a binational partnership for scientific assessment and data collection on transboundary aquifers.[11] The IBWC played a central role in developing the TAAP by designing a joint process by which the US and Mexico could cooperate at the national level. The joint process limits TAAP activities to those that benefit both countries and requires that they neither infringe on the legal frameworks of individual nations nor contradict existing binational agreements (IBWC, 2009).

The main aim of the TAAP is to improve transboundary-groundwater knowledge through scientific assessment via binational, aquifer-specific technical committees (Megdal & Petersen-Perlman, 2018). Yet, in addition to data-collection and information-sharing advances, the TAAP has also helped build cross-border relationships and promote community outreach (Megdal & Scott, 2011). Organizations on both sides of the border – the Universidad de Sonora, Comisión Nacional del Agua, USGS, and University of Arizona – have engaged in activities including binational hydrogeologic mapping and modelling (via technical meetings), stakeholder workshops, site visits, multi-state meetings and binational summits. A binational, bilingual Santa Cruz assessment report is expected in 2019.[12] In addition to the TAAP, numerous other efforts in the Upper Santa Cruz basin have benefited from collaboration among university researchers and NGOs.

The Upper Santa Cruz case illustrates how US–Mexico asymmetries in economic development, infrastructure, legal-political structures, and implementation of environmental policies, combined with groundwater dependence and climate change uncertainty, complicate water security. Nevertheless, both formal agreements and less formal cooperative scientific efforts have been achieved. Efforts in this basin demonstrate the value of (1) addressing resource challenges through an existing international commission, and (2) focusing initially on technical aspects that avoid sovereignty concerns while productively addressing information gaps that are key to water-security challenges. This is one approach to enhancing transboundary water security even in the face of political barriers – other approaches may better address other types of water-security challenges, as we see in the next case.

Case study 2. Water security and desalination in the US–Mexico border region: binational implications of water transfers

Binational desalination – that is, desalination that takes place on one side of the border, but benefits both sides either directly or indirectly – is seen by some as a potential solution to water scarcity for the US–Mexico border region (Wilder et al., 2016). This case examines the water-security implications of proposed desalination facilities that would be located on the northern coast of Sonora, Mexico, but provide water supplies for both Mexico and the US. Using the water-security-prism framing (Figure 1), concerns about the region's fragile *ecosystems* and insufficient water *quantity* come to the fore. It also reveals how *transport* and *energy* requirements for desalination elevate the cost of water, affecting water *access*. And within the existing complex institutional setting and volatile *political* environment, effective environmental *policies* are needed to ensure a balance of benefits for both sides.

Water-security challenges

The US–Mexico border region is subject to environmental, demographic and economic forces that compromise water security (Leichenko & O'Brien, 2008). In terms of environmental forces, water scarcity is heightened by climate variability and change, and by overexploitation of the Colorado River, a major transboundary river, and aquifers. In the semiarid Sonoran Desert, rivers and creeks are intermittent or ephemeral, and precipitation is scant (Table 2). Climate projections for the western US–Mexico border region in particular – combined with social and political forces – suggest likely rises in water scarcity and water demand (Garfin et al., 2014). Economic growth, including crop agriculture, ranching, mining, manufacturing and urbanization, along with associated population growth in the border region, further increases water demand (Wilder et al., 2012). Specifically, the Upper Gulf of California (or Sea of Cortés) is a semi-enclosed sea that has been heavily impacted for decades by lack of inflow from its main source, the Colorado River (Kennedy, Rodríguez-Burgueño, & Ramírez-Hernández, 2017). In addition, groundwater has been the main water source for decades, and aquifers throughout the Sonoran Desert region are at the brink of depletion or are already depleted (Palma, González, & Cruickshank, 2015).

Influence of the transboundary context

This transboundary region exhibits an uneven distribution of populated areas and multiple jurisdictions with different water-management approaches. As mentioned in the previous case study, this region – divided by the US–Mexico international border – features different languages, cultures, legal traditions, forms of government and levels of development on either side. The borderline cuts through the Sonoran Desert ecosystem and the historical territories of indigenous groups, like the Tohono O'odham Nation. As noted above, the neighbouring nations pursue very different water-management and governance approaches. In such a setting, binational issues typically have been handled through nation-to-nation approaches.

As water managers in both countries search for additional water supplies that can sustain an increasing population under projected water shortages, desalination has emerged (for some) as an attractive, 'drought-proof', water-security-enhancing option (Hirt, Snyder, Hester, & Larson, 2017). Recently some local steps and agreements have been made towards this end. In 2012, for example, IBWC signed an agreement on research exploring two prospective desalination plants (Figure 4). The first would be in Rosarito, Baja California, along Mexico's Pacific coast; and the second in Puerto Peñasco, Sonora, on the Sea of Cortés (IBWC, 2012).

A desalination plant in Rosarito could deliver desalted water to the city of San Diego in California (Dibble, 2014). A plant in Puerto Peñasco could provide desalted water for Arizona and possibly Nevada, depending on the facility's scale (Wilder et al., 2016). An alternative binational desalination scheme relies on the US investing in a desalination plant in Mexico in exchange for water rights from the Colorado River. However, in spite of the benefits offered by these two binational projects, the prospects for their actual construction remain unclear. Mexico's CONAGUA would be the ultimate arbiter for such projects, but it has remained silent on them, the main apparent obstacle to approval (Wilder et al., 2016).

In addition to securing CONAGUA's approval, all such initiatives must navigate a maze of legal and administrative challenges. Several reasons render these efforts politically sensitive, and they are viewed by many officials as a losing proposition, with the US reaping the lion's share of the benefits (more water), and Mexico bearing the brunt of the costs, especially environmental externalities associated with desalination. A desalination plant such as the one proposed for Puerto Peñasco would likely affect commercial fishing and sensitive ecosystems that are currently designated as a marine reserve (Figure 4) (Hirt et al., 2017; Ocasiot, 2015). The primary by-product of desalination – a brine that is deposited into the sea – alters the salinity and temperature of the coastal marine ecosystems (Gleick, 2015; Ocasiot, 2015). The desalination technology also employs chemicals that harm some marine species, and the intake process inevitably catches fish (Gleick, 2015). If desalted water is conveyed to the US, terrestrial ecosystems could also be negatively affected by the conveyance system. Pipes would need to cross a protected area, the Pinacate y Gran Desierto de Altar Biosphere Reserve, which lies between Puerto Peñasco and Arizona, fragmenting the reserve and disturbing the habitat of many terrestrial species (Wilder et al., 2016).

Institutional capacity and transboundary cooperation

Inadequate environmental regulations for monitoring and handling brine disposal in Mexico exacerbate the consequences of ecological degradation due to desalination. The Ministry of Natural Resources and the Environment (SEMARNAT) does require an environmental impact assessment for the construction of a desalination plant, and CONAGUA would need to issue a discharge permit. But beyond these two requirements, there exists no regulation for brine discharges (McEvoy, 2015). Environmental regulations for desalination are urgently needed in Mexico to protect ecosystems and

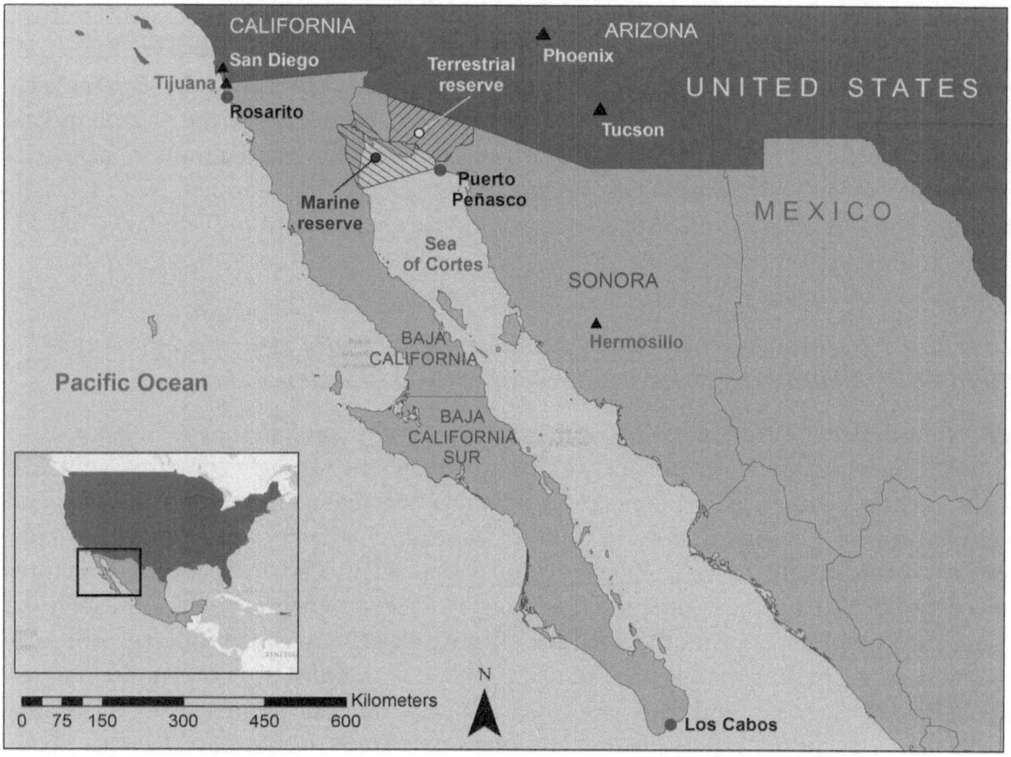

Figure 4. Location of desalination plants and marine and terrestrial reserves in the Sonoran Desert. Basemap sources: Esri, DeLorme, HERE, MapmyIndia.

related livelihoods (Pombo, Breceda, & Valdez Aragón, 2008). Other important factors for binational desalination are the huge energy requirement and the greenhouse-gas emissions associated with the process (Meerganz von Medeazza, 2005). The use of renewable sources is not currently feasible for large-scale energy supply (Wilder et al., 2016), which constrains cost–benefit ratios.

Should binational desalination at these sites or elsewhere become a reality, US cities undoubtedly would be the largest consumers of the desalinated water. But such projects also can be attractive to Mexico, because additional water would allow Mexican cities to grow and develop, while also providing local employment. What remains uncertain, however, is how Mexican enterprises and residents would pay for desalinated water that would cost seven times what they pay now (Wilder et al., 2016). The desalination plant in Los Cabos, Baja California, offers a glimpse of what could happen to make water prices affordable for the local residents: the city highly subsidizes the desalted water, absorbing the high cost of production – but with no sustainable future for business operation (McEvoy, 2014).

Although desalination shows promise for increasing supply, it has a steep price. Binational desalination that creates disproportionate negative externalities for Mexico could enhance disparities and aggravate binational political tensions (Hirt et al., 2017). Elucidating the various water-security attributes of this case reveals that improving water security for the transboundary Sonoran Desert and Upper Gulf of California region will require multiple conditions: technical and environmental feasibility of the binational desalination project and its conveyance system; new, more favourable economic conditions, including cheaper renewable sources of energy; a more responsive and innovative institutional framework that includes monitoring of the environmental impacts of desalination; and a more integrative, cooperative and mutually beneficial binational political relationship. In the next case, we examine another pair of neighbouring nations, Ecuador and Peru, where political relationships are similarly influential in transboundary water security.

Case study 3. Water security in the Ecuador–Peru border region's Catamayo-Chira basin: limitations of local coordination in a binational river basin

The transboundary Catamayo-Chira basin crosses the mountainous Ecuador–Peru border (Figure 5). In both nations, the basin's residents struggle to access sufficient water for their daily needs. Cooperation on water management between Ecuador and Peru, which have a long history of conflictual relations, has evolved slowly but consistently (Toledo, 2017). Still, differences in management and institutional structures make it difficult to address binational water challenges. Viewing this case through the water-security prism exposes the need for effective *policies* and international coordination – i.e., promoting peace and assuring *political stability* – between the foreign ministries of Ecuador and Peru to close the gap in transboundary water-security challenges – primarily water *quality, sanitation and access* – that have for years been addressed mostly locally.

Water-security challenges

Water-security challenges in the Catamayo-Chira basin are driven by high climatic variability (e.g., the El Niño Southern Oscillation) and anthropogenic impacts on water quality, which contribute to insufficient access to drinking water and uncertainty for local livelihoods, which rely on irrigated agriculture. Access to potable water is a perpetual challenge – only 59% of basin inhabitants have potable water service (PNUD, 2015). Dumping of untreated wastewater from domestic, agricultural, industrial and mining activities degrades surface water quality (ANA, 2015). Although considerable aquifer storage is estimated, groundwater use is limited (AECID, 2005). To compound water-supply and quality challenges, service providers are generally unreliable, and potable-water management remains weak. Due to increased rainfall variability, surface water storage reservoirs are becoming less reliable. On the Peruvian side, the Poechos reservoir maintains less than 50% of its original capacity (which was 1000 Mm^3) due to sedimentation. Traditional practices – such as *tajamares* (protective walls) and *albarradas* (artificial, earthen-dike wetlands) – through which small dams store water, have been abandoned and not yet replaced with functional management regimes (PNUD, 2015).

Influence of the transboundary context

On the national level, Ecuador and Peru have their own institutional structures to guide water governance (Table 4). However, these arrangements differ between the two countries, and both management systems and regulations have been inconsistent over

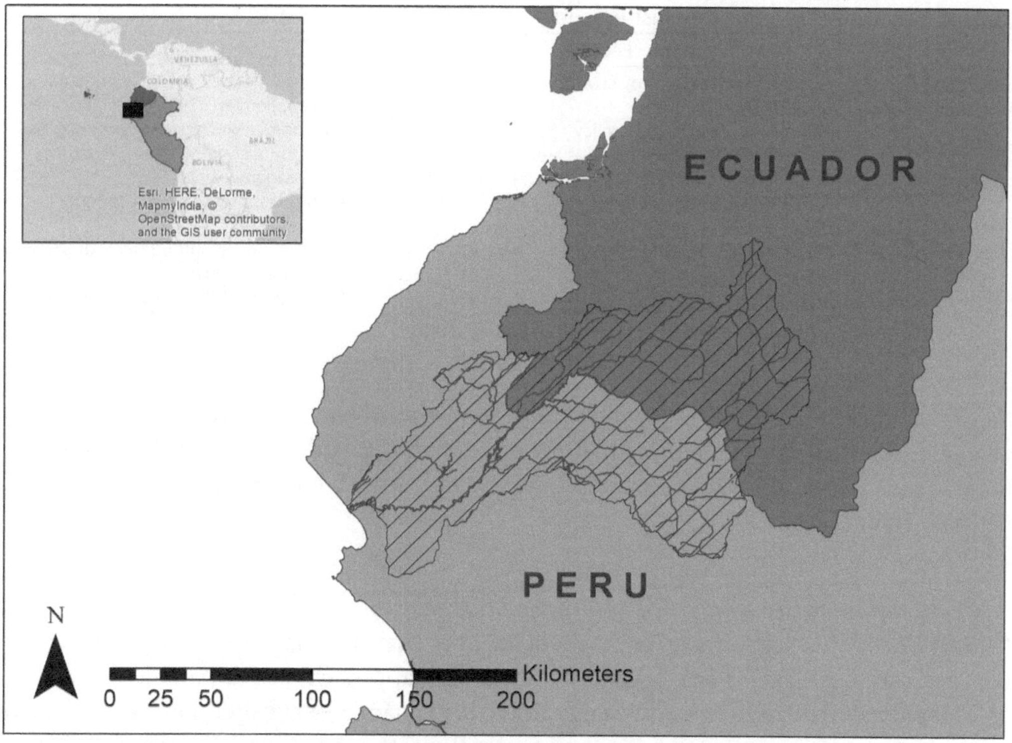

Figure 5. Location of Loja Province, Ecuador, and Piura Province, Peru, and the Catamayo-Chira River basin. Basemap sources: Esri, DeLorme, HERE, Mapmylndia.

time. While in Peru a water law was approved in 2009 that included specifications for water management in transboundary basins (motivated in part by Peru's goal of rising to the economic level of OECD full-member countries), in Ecuador a new water law was not approved until 2014. During the absence of specific legal and institutional frameworks in both nations – according to some Peruvian technicians involved – dialogues regarding Integrated Water Resources Management (IWRM) in transboundary basins were infrequent.

Ecuador and Peru had a sustained period of border conflict and political crisis, from their early-nineteenth-century beginnings until 1998. However, as in other border regions, political tensions were often transcended by strong cultural kinship and high social mobility between border communities, which increased economic transactions and strengthened cultural connections. In spite of political conflict, the international border did not prevent the circulation of ideas, people and products (Hocquenghem & Durt, 2002).

The 1998 peace agreement and other efforts towards international cooperation – such as the Catamayo-Chira Binational Project, promoted by the Spanish Agency for International Development Cooperation (AECID) – have deeply influenced local communities thanks to direct-intervention strategies in border areas. One of this project's outcomes was the 2011 Land Use Planning, Management and Development Plan for the Catamayo-Chira transboundary basin, which was recognized by local authorities in Loja (Ecuador) and Piura (Peru) but not endorsed by other government levels. Nevertheless,

Table 4. National-level regulatory instruments in the Catamayo-Chira basin.

Instrument	Ecuador	Peru
Environmental management	Decentralized System of Environmental Management (2004) National Environmental Policy (2011)	National System of Environmental Management (2004) Environmental National Policy (2009)
Water management	Strategic National System of Water (2014) Organic Law of Water Resources, Use and Exploitation (2014) and bylaw (2015)	National System of Water Resources Management (2009) Water Resources Law (2009) and bylaw (2010)
Water-resources policy	National Plan of Water (2016)	National Policy and Strategy of Water Resources (2015) State Policy of Water Resources (2012)

Source: Water Secretariat (Ecuador) and Water National Authority (Peru).

the local governments of Loja and Piura have spearheaded efforts at cross-border dialogue, sidestepping mediations that are typically regulated by the two foreign-affairs ministries. Such local-level actions typically are less sensitive to sovereignty issues than national-level dialogues.

Institutional capacity and transboundary cooperation

Institutionally, both countries have changed how they manage their water resources. For example, in 2009 Peru adopted an IWRM approach accompanied by a new organizational architecture; comparably, in 2014, Ecuador established new institutional practices for water-resources management. In Peru, the Autoridad Nacional del Agua (ANA, the National Water Authority) is responsible for managing water resources and serves as the highest technical and regulatory authority in the National System of Water Resources Management. As in many countries with important agricultural sectors, this institution is housed within the agricultural ministry. ANA coordinates with the foreign-affairs ministry when signing multinational agreements for integrated water management in transboundary basins. In the structure of Peru's new water-resources law, the Sub-directorate of Transboundary International Waters was created in the foreign-affairs ministry. This agency executes and evaluates foreign policy to safeguard Peru's transboundary-waters interests. Thus, Peruvian law assigns ultimate authority for management of these interests to its foreign-affairs ministry, with technical participation from ANA.

In Ecuador, the National Water Secretariat (SENAGUA) is responsible for guiding water-resources management in an integrated and sustainable way using an ecosystem approach. Created in 2008, SENAGUA operates directly under the Presidency of the Ecuadorian Republic, with its own resources and budget, and with technical, operational, administrative and financial independence. As in Peru, the Ecuadorian Ministry of Foreign Affairs and Mobility shares the authority to establish binational-cooperation guidelines to maintain proper functioning of the Binational Development Plan. This ministry is a member of the binational Ecuador-Peru Neighbourhood Commission, which promotes and coordinates programmes, projects and activities according to the interests of both countries.

In addition to the national ministries, extra-governmental actors (e.g., USAID, European Union, Spanish AECID, German GIZ, and Swiss SDC) often play an important role in moderating bilateral water-resources governance in the Catamayo-Chira basin via local- and national-scale projects. Examples include providing seed funds to implement basin-level water-resources management, and promoting binational dialogues whenever appropriate.

Some regional institutions – such as organizations, policies, binational strategies, and agreements – exist between Ecuador and Peru that help guide transboundary water-resources management (PNUD, 2015; Binational Development Plan, 2016; Toledo, 2017):

- Presidential Statements between Ecuador and Peru (2007, 2009, 2010, 2012–2017)
- Proposed Binational Commission for IWRM between Ecuador and Peru (2015)
- Quadripartite agreement between ANA (National Water Authority, Peru)–SENAGUA (National Water Secretariat, Ecuador)–Binational Plan–Binational Fund Ecuador Peru (2014)
- Andean Strategy for IWRM in the frame of the Andean Commission of Nations (2011)
- Binational Plan for Border Region's Development (2007)
- Peru-Ecuador Comprehensive Border Integration, Development and Neighbourhood Agreement (1998).

The presence of institutional frameworks and legal water-management tools offers the potential to encourage and enhance binational dialogues and actions in the Catamayo-Chira basin. Yet, current policies and agreements do not yet provide a clear basis for binational cooperation, and national-sovereignty concerns between Ecuador and Peru often prevail over environmental ones. Nevertheless, according to basin water-management representatives, similarities between the two nations are more common than differences, especially regarding water quality and ecological conservation.

The possibility – even if it remains remote – of achieving consensus on water issues that affect both sides of the border may represent the limit of bottom-up processes. The Catamayo-Chira Binational Project illustrates the challenge of translating local-level capacity into higher-level (e.g., national or binational) understanding. In this basin, water availability, quality and access clearly are difficult to secure in view of physical conditions and anthropogenic impacts – in addition to national-sovereignty sensitivities. But the recent design of tailored, cooperative institutional frameworks, some of them binational, offers hope for overcoming those barriers and solving mutual problems jointly and equitably. Designing legal frameworks and policies to promote equitable water security can be a challenge, even within a single nation, as the next case demonstrates.

Case study 4. Water security in Peru's subnational Ica River basin: transjurisdictional asymmetries between poor upstream and prosperous downstream provinces

The Ica River basin extends across two departments in Peru,[13] and thus represents a *transjurisdictional*, transboundary context situated entirely within one country (Figure 6). In spite of a single political system and shared language and culture, multiple water security challenges are revealed via the water-security prism. These stem from the Ica River basin's extreme disparities between the two departments – most notably, in water *quality, economic growth, sanitation, political stability/power, transport,* and water-resource *access.* This basin presents an atypical upstream–downstream relationship, where downstream water users hold more economic and political power than those upstream. The Peruvian government is planning to expand interbasin transfers to further supply the downstream, coastal communities, but these plans have spurred regional conflict.

Water-security challenges

Water-security challenges characterize the entire basin, though they are very different in coastal and highland areas. In the coastal lowlands around the city of Ica, the agricultural (e.g., for export) and industrial sectors rely heavily on groundwater, which has declined rapidly. As new areas of reclaimed desert are converted to agricultural production, thousands of deep wells are being drilled in the valley. Many smallholder wells and urban wells have gone dry, leaving suburban and rural families without access to water. Extreme responses have included corruption among water users and regulators, 'land grabbing' for water-rights access, efforts to use surface water to recharge the aquifer, and now pressure to expand the existing interbasin Choclococha transfer project – a system of tunnels, reservoirs, and canals – to draw more Amazon basin water to the dry Pacific Ica basin (Cárdenas Panduro, 2012; Damonte, Pacheco, & Grados, 2014; Urteaga, 2014). Water quality has been degraded by urban and industrial waste from

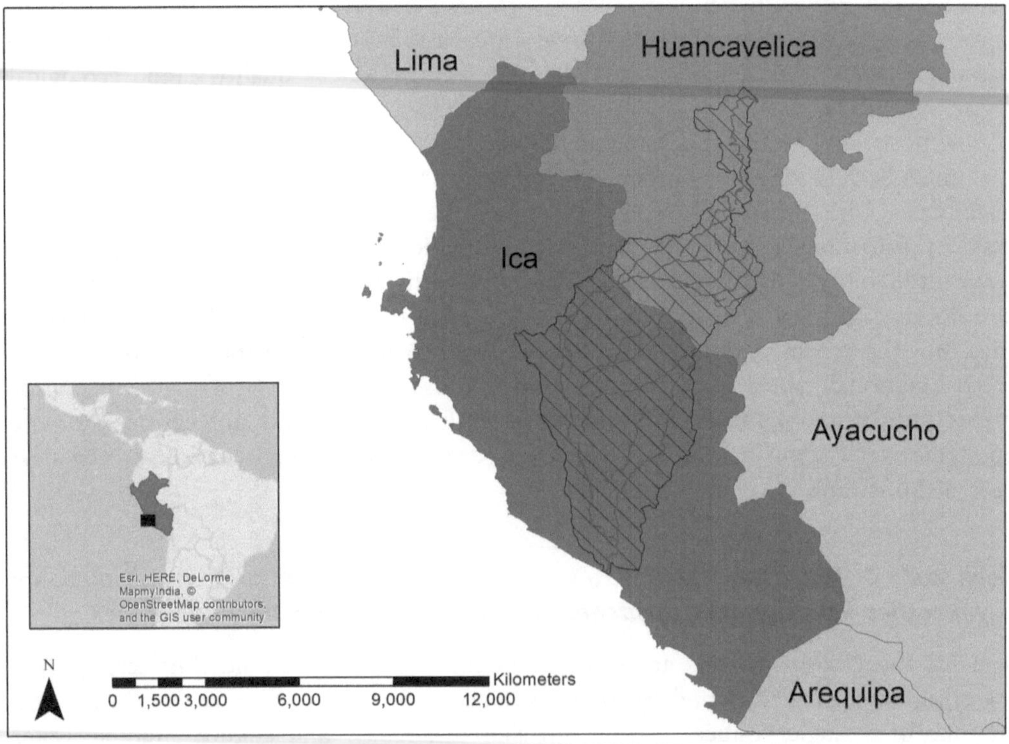

Figure 6. Location of the Ica River basin in central Peru. Basemap sources: Esri, DeLorme, HERE, MapmyIndia.

intensive agriculture, expanding urban and peri-urban centres and lack of public facilities, and highly permeable desert soils.

In the intermountain and the high-alpine *puna* regions, water insecurity presents another paradox: plenty of rain falls, but water shortages may remain because precipitation is spatially and temporally uneven, and because surface water may be highly contaminated by livestock, human waste, and mining activity. In a survey of Ica basin residents in the department of Huancavelica, Herz Sáenz and Huamán Arias (2014) found that 74.5% of interviewees experienced consistent water scarcity, and 16.4% said that water was scarce during some times of the year and/or in some locations. Many families struggle to provide water for irrigation and livestock, and most do not have basic water services – either access to potable water nearby or septic facilities (INEI, 2007). Fuel to boil water is scarce, and access to medical facilities to treat water-borne diseases is often unavailable.

Influence of the transboundary context

Intra-national, transjurisdictional challenges – e.g., between neighbouring states or provinces – are highlighted in the Choclococha interbasin transfer system, completed by the Peruvian government in 1958 to move water from the Andes-Atlantic Pampas basin to the Andes-Pacific Ica basin. In Peru, the Amazon basin receives 90% of the water, but 90% of population is in the arid Pacific basin. By capturing Amazonian-slope water and transporting it via a mountain tunnel into the Ica River to flow for some 320 km to the Pacific coast, the Choclococha project expanded irrigation in the Pacific coastal lowlands of the valley by 10,000 hectares. Along the flow path, in the Tambo intermountain region, Choclococha water has extended available canal water by several months, though farmers in other tributaries still rely heavily on rainfall and springs. Approximately two-fifths of the Choclococha water is lost to infiltration, evaporation and use before it reaches the Ica, accelerating demand for still more water to be diverted from the highlands and increasing political, environmental and social conflicts (PETACC, 2010).

The prospect of a Choclococha expansion has generated tension among residents of both Huancavelica and Ica, as national-government-funded projects draw even more Andean highlands and Amazonian Pampas basin water for the needs of coastal agriculture, industry, and urban expansion. Although both Ica and Huancavelica are geographically involved in this project, thanks to its superior economic power and prominence in federal politics, the Ica regional government downstream determines the legal and political outcomes of water-resource decisions (Damonte Valencia, 2015; Geng, 2016). While Huancavelica's rural and isolated communities upstream have participated actively in the policy process and in integrated-policy dialogues, their influence in water distribution and use has been very limited.

Water demand in Ica has fueled myths, tensions, inequity in representation, and conflicts at all scales, such as altercations between small farmers and agro-exportation companies, domestic and agricultural users, and Ica and Huancavelica for their respective water rights (Boelens & Vos, 2012; Hepworth, Postigo, Güemes Delgado, & Kjell, 2010; Herz Sáenz & Huamán Arias, 2014). A common perception – based on our experience in science-policy dialogues and stakeholder meetings – is that Huancavelica is water-rich, and water is either not needed (because the region is too cold to grow crops), used inefficiently, or not as necessary as in Ica. Lima-based NGOs and politicians have exacerbated the dichotomy for political reasons, either to integrate this case as a component of environmental/conservation strategies or as part of a larger political agenda, primarily in the context of state elections. Huancavelica residents have received only small advantages from the interbasin transfer (e.g., locals are employed seasonally to clean the canals) while losing a large part of their water resource. At the same time, they maintain that the effects of the diverted water on downstream communities in the Pampas basin, on the eastern Andes slope, remain unexamined.

Institutional capacity and transboundary cooperation

Currently, Ica and Huancavelica state agents and NGOs are engaging in natural-resources management dialogues to resolve water conflicts. Ica, with its economic advantage and political power, dominates the debate. The Peruvian government has exercised contradictory roles in the process, taking advantage of dialogue progress to push for the extension of the Choclococha system while also supporting other minor regional development projects and environmental agreements. The decision-makers remain largely those with political sway – local water authorities from both states have participated in these discussions, but have not significantly influenced the outcomes (Geng, 2016).

Across the Andean gradient, mountain-based mining and hydroelectric generation and lowland agro-export industry support a thriving national economy and enhance the communities' resource security. Still, resource challenges and competition have contributed to environmental degradation, social deprivation, and conflict. Additional drivers directly and indirectly impact farmers' and consumers' livelihoods.[14] NGO and government initiatives have promoted idea exchanges between Ica and Huancavelica, but the thriving coastal economic engine still holds sway. In such a setting, inadequate water security and insufficient transboundary discourse could worsen environmental, economic and social outcomes in the Ica Valley.

The Peruvian inter-state, transboundary Ica–Huancavelica example illustrates the potential for *intra-national* tensions, even where participants share common nationality, languages, culture, laws, and political frameworks for resolving water conflicts. The degree of mutual animosity and conflict in Peru underscores the even greater difficulty of achieving *international* transboundary agreements, which may not offer the collective domestic benefit needed for water-management solutions. The next case demonstrates the challenges of achieving international coordination for regional collective benefits when water security is complicated by uncertain future conditions and water sources lie far removed from residents' everyday lives.

Case study 5. Water security in the Andes: climate change and transboundary glacial headwaters in the Chile–Argentina border area

Meeting in the Andes mountains at the international divide, the Maipo and Mendoza River basins share a common origin in the high-elevation mountains (Figure 7). Downstream urban areas – the Chilean capital of Santiago and the Argentinian city of Mendoza – both depend on snowmelt and glaciers that traverse the Chile–Argentina international border as a chief component of their water supplies.

This case highlights an unusual transboundary context, yet one that is sure to become more important in areas where climate change threatens the high-mountain glaciers and seasonal snowpack that supply water for users downstream. Viewed through the prism of water security, this case sheds light on the *uncertainty* involved in water-security challenges driven by *global change* and the potential conflict between preserving glacial water sources – which provide water *quantity* – and advancing the supply of *industrial resources*. In Chile and Argentina, government *policies* and private-sector activities regarding Andean glaciers affect both nations' water security.

Water-security challenges

Due to its relatively low precipitation, this transboundary region supports agriculture only with irrigation. In Mendoza, agricultural systems, mainly vineyards, olive groves, fruit orchards and vegetable farms, form a growing oasis (Abraham & Villalba, 2008). New agricultural projects, along with residential and touristic developments, have stressed existing surface water distribution systems and increased reliance on groundwater. Current irrigation practices – surface application of large volumes of water – are inefficient and degrade soil and groundwater quality.

Over the last few decades, increasing urban development and commercial expansion (e.g., petroleum exploitation and tourism) in the Mendoza basin have raised water demands on local aquifers (Hurlbert, 2018). The situation in the Maipo basin is similar,

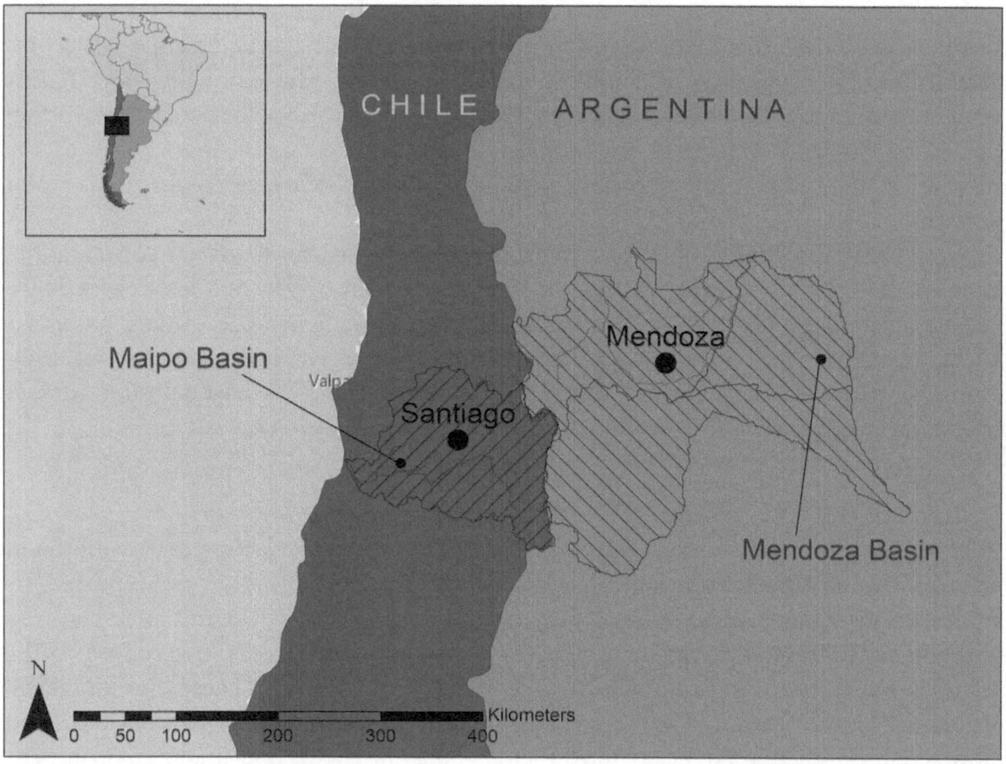

Figure 7. Location of the Maipo River (Chile) and Mendoza River (Argentina) basins. Basemap sources: Esri, DeLorme, HERE, MapmyIndia.

but in this basin (which covers twice the area of the Mendoza basin) agriculture and industrial activities (e.g., mining) must coexist with the major urban centre of Santiago. In the upper reaches of both basins, high-elevation open-pit mines pose a risk of pollution, and are often below glaciers – key water supplies for downstream areas (Khadim, 2016).

Both basins experience interannual hydroclimatic variability (alternating wet and dry years) associated with the El Niño Southern Oscillation (Aceituno, 1989). Long-term trends suggest that reductions in snow accumulation and accelerated glacier melting will likely reduce basin-wide water availability (Masiokas, Villalba, Luckman, Le Quesne, & Aravena, 2006). More winter precipitation will fall as rain, versus snow (Surfleet & Tullos, 2013), causing 'rain-on-snow' events that can reduce glacier water storage and cause earlier and faster melting periods. These changes could also prompt severe flash flooding in the mountain foothills.

The lower water availability and rising water demands will likely affect the reliability of water-use rights for agriculture differently in each basin (Meza, Wilks, Gurovich, & Bambach, 2012). This is because water rights in Mendoza are tied to the land, whereas in Chile water rights can be obtained via trades, independent of land ownership. In this context, water-management schemes that are adaptive and use alternatives for supply and demand management are becoming critical (Richardson, Steffen, & Liverman, 2011).

Influence of the transboundary context

In this transboundary region, the Maipo and Mendoza basins are linked via trade and economic activities. The route connecting Mendoza to Santiago is an important commercial link that transports five million tonnes of merchandise annually. Both basins largely rely on agricultural and related activities (e.g., vineyard-associated tourism) that depend on Andean snowmelt. In addition to agriculture, a key contributor to the national GDP of both Chile and Argentina is mining, which directly threatens water resources.

While Chile and Argentina have taken individual approaches to glacier conservation, a coordinated, transboundary strategy would help ensure preservation of these high-mountain supplies throughout the region. Glaciers, which provide critical baseflows during drought, are considered priority water-security resources because of their imminent threat of depletion. In contrast, the importance of seasonal snowpack is sometimes overlooked and considered an easily replenished resource.

Institutional capacity and transboundary cooperation

Chile and Argentina have advanced differently with regard to glacier conservation and management. Regional advocacy for glacier-conservation legislation was spurred by large-scale, open-pit Andean mining projects in the early 2000s. The initiative began as local efforts to preserve indigenous peoples' water and land access in response to the Barrick Gold Corporation's Pascua-Lama and Veladero projects. These developments, located in high-elevation Andean crests in Chile and Argentina, incited local and national environmental advocacy movements (Talliant, 2013). Social and environmental activist groups organized protests, road blockages and publicity campaigns, and lodged formal complaints with the OECD (Khadim, 2016).

In Argentina, these environmental movements garnered the support of NGOs, researchers and political parties to promote a glacier-protection law in 2008 (Law 26.418; Scott et al., 2012b). While the law was initially vetoed due to mining-sector pressure (Isla Raffaele, 2016), its promoters launched a major national campaign denouncing the sector's complicity with the national executive power and raising awareness of the importance of protecting the sources of watersheds (Isla Raffaele, 2016; Wagner, 2017). Finally, in 2010, the glacier law was sanctioned (Law 26.639, 2010). The Argentinian National Glacier Act (Talliant, 2013) prohibits activities that could alter the natural condition of glaciers, including mining and oil exploration; protects freshwater and biodiversity; and mandates a national glacier inventory (Khadim, 2016), be completed in 2017. Mining is also legally regulated in the province of Mendoza (Law 7722, 2007).

However, while the Argentine national government has a mandate to prevent pollution, provinces have original jurisdiction over natural resources – thus, controversy exists regarding who can approve mining projects (Talliant, 2013). The ongoing power struggles between the Argentinian national government and provinces – which might benefit from permitting economic development – reveal a limitation of federal policy-making on this issue.

Environmental organizations and indigenous groups have staged opposition to mining expansion in high-mountain areas in Chile, as well. While local opposition movements have been supported by national environmental NGOs, Chilean government support for these movements is limited. Local governments typically side with mining companies to support economic development, and the national government has failed to pass comprehensive legislation or to address the petitions of opposition groups (Urkidi, 2010). While limited discussion on this issue continues, a national law for glacier conservation does not appear to be on the horizon.

Although there have been a few formal agreements between the two countries – one example is the Treaty on the Environment and the Additional Protocol on Shared Water Resources (1991) – binational channels established by environmentalists and scientists have been much more prevalent than channels at the state level. Researchers at IANIGLA[15] and other Argentine scientific institutions cooperate regularly with Chilean counterparts, as well as with a variety of environmentalists. IANIGLA is leading an effort to map and catalogue glacier formations.

Recently, due to a series of cyanide-laden spills at the Veladero mine in Argentina's San Juan Province, environmental activists sued IANIGLA scientists for not inventorying glacier formations smaller than one hectare.[16] The scientists argue that it is not methodologically feasible to delineate such minor formations and that their contribution to runoff is insignificant. This approach garnered criticism from environmental groups, who claimed that IANIGLA was not doing enough to implement the full scope of the glacier-protection law. This case has thus become a great 'socio-technical-*cum*-legal' controversy that adds enormous complexity and foreshadows a difficult resolution. Institutional arrangements are needed that help align scientific approaches with social and environmental concerns.

It appears evident that the Maipo and Mendoza River basins, though they are in different nations, might have to prepare jointly for a water-secure future. Among the steps needed to assure snow and ice protection in the Andes are: adaptive operating rules for water-distribution systems; increasing investments in low-cost technologies that promote enhanced water-use efficiency and equitable distribution; and development of dedicated, coordinated glacier-protection programmes. These steps would increase resilience to climate change and strengthen water security in this important trans-Andean region.

Discussion: water security and institutional responses in transboundary contexts

Examining transboundary water issues through the water-security prism helps unravel the multiple and varied attributes of transboundary water challenges in the arid Americas. The challenges identified reflect the multidimensional and multi-attribute nature of water security, and how transboundary contexts may further confound already complex challenges. Our study also highlights how institutional responses were employed – or are still needed – in each case. In Table 5, we summarize the key water-security attributes, transboundary nature, water-security challenges, institutional capacities and legal frameworks for our five case studies.

Overall, in terms of water-security challenges, we find that in arid environments, water quantity and quality and the uncertainties of climatic changes are common concerns. The need to protect existing water sources from pollution – whether caused by industrial sources or a lack of adequate wastewater infrastructure – came to the fore in multiple cases (e.g., the Upper Santa Cruz, Catamayo-Chira and Ica basins). Climate-change drivers of water insecurity were particularly relevant for groundwater and glacial-water sources (e.g., Upper Santa Cruz, Andean glaciers). For example, rising water demands coupled with likely reductions in groundwater recharge in the US Southwest (Meixner et al., 2016) may make groundwater dependence in the Upper Santa Cruz basin unsustainable. Climate-change impacts on glaciers are expected to be significant – worldwide, glacier retreat is expected to reach 60% by 2050 (Bates, Kundzewicz, Wu, & Palutikof, 2008) – a concern for shared Andean glacial stores.

Our cases also exhibited distinctive water-security challenges. Some cases focused more on water supplies – either emphasizing conservation of existing sources (e.g., the Andes glaciers) or expanding supplies via technology (e.g., binational desalination) and interbasin transfers (e.g., the Ica basin). The negative impacts on ecosystems and the cost of energy needed to desalinate and transport water supplies is a concern specific in the Northwest Mexico case (e.g., binational desalination). Access to potable water was a salient feature of both the Ica and the Catamayo-Chira River basin cases. Our analysis also identified issues related to the water–energy–food nexus (e.g., Ica basin and Andean glacier cases) and the water–energy nexus (e.g., binational desalination). Still other water-security attributes revealed in the case studies reflect their transboundary nature, as we discuss below.

How are issues of water security affected by transboundary conditions?

Our case studies demonstrate that almost invariably, the presence of borders compromises water security in arid regions. Although we discover that many factors compound the challenge of addressing water security in transboundary contexts, particularly

Table 5. Key observations of five case studies.

Site and location	Key water-security attributes	Transboundary nature	Water-security Challenges	Institutional capacity	Legal frameworks
Upper Santa Cruz basin, Arizona–Sonora, US–Mexico border	• quality • sanitation • global change • uncertainty • economic growth • political stability	Binational aquifer (and basin)	Cooperation on transboundary aquifer assessment amid contentious binational relations	Transboundary Aquifer Assessment Program (science-based); International Boundary and Water Commission; national water commission (Mexico); state water management agencies (US)	1944 Water Treaty and La Paz Agreement limited for groundwater
Northwest Mexico Desalination, Sonora and Baja California, Mexico	• quantity • ecosystems • energy • transport • policy • political stability	Binational desalination and water transfers	Binational implications of water transfers from desalination plants in fragile ecosystem and volatile political environment	International Boundary and Water Commission; national water management and environmental agencies (Mexico)	Limited environmental regulations in Mexico; underdeveloped framework for binational benefit-sharing
Catamayo-Chira River Basin, Ecuador–Peru border	• quality • sanitation • access • policy • political stability	Binational river basin	Institutional arrangements to govern transboundary rivers amid international mistrust and dispute	Andean Commission of Nations; Binational Commission for IWRM; Binational Plan for Border Region Development; Peru-Ecuador Comprehensive Border Integration, Development and Neighbourhood Agreement; national water policies	Peru water law (2009); Ecuador water law (2014); binational peace agreement
Ica River Basin, south-central Peru	• quality • sanitation • access • economic growth • transport • political stability	Subnational, transjurisdictional river basin	Transjurisdictional asymmetries between poor upstream and prosperous downstream provinces	Ica regional government; state agents and NGOs	Peru national water law (2009)
Maipo River Basin (Chile) and Mendoza River Basin (Argentina) – Andean glaciers	• quantity • global change • industrial resources • policy • uncertainty	Binational glacial headwaters	Changing water availability from transboundary sources of ice and snowmelt	National water agencies; civil society groups; scientific institutions (IANIGLA)	Argentina's National Glacier Act; Treaty on the Environment and Additional Protocol on Shared Water Resources (limited inclusion of glaciers)

problematic conditions are traditions of political contention and cross-border asymmetries that inhibit coordinated institutional responses.

The Catamayo-Chira River basin in Ecuador and Peru offers the starkest contrast between neighbouring countries. Fueled by a history of contentious international relations, Ecuador and Peru developed strategies for water management by different methods on vastly different timelines. As a result, their respective institutional arrangements for managing water do not align well. The tense political history between these nations has heightened concern about each nation's individual sovereignty, challenging attempts towards faster international negotiation. As noted by long-time water-conflict scholar Aaron Wolf (2007), such imbalances in growth and institutional capacity, accompanied by a tradition of contentious relations, confound international cooperation over shared water.

National-sovereignty concerns and incongruous institutional arrangements for water governance also confound water security between the US and Mexico. Cross-border coordination is more difficult due to differences between the centralized Mexican system and the decentralized, often fragmented, US system of water governance.

Information asymmetries also persist across borders, particularly when dealing with hard-to-measure groundwater resources. Sufficient information is not always available to both sides because nations and agencies may be unwilling to share it, or the appropriate channels for sharing it do not exist, or the extent, type and quality of the collected data differ (Linton & Brooks, 2011; Milman & Ray, 2011). Accordingly, binational efforts on transboundary groundwater in the Upper Santa Cruz basin (e.g., the TAAP) have focused on addressing information asymmetries through scientific assessment, while simultaneously avoiding national-sovereignty concerns.

Moreover, economic, social and cultural asymmetries – and their effect on binational power dynamics – emerge as complicating attributes for transboundary water security. Along the US–Mexico border, Mexico has experienced faster industrial and urban growth in the past few decades than the US. In the Upper Santa Cruz basin, this trend has resulted in asymmetric water needs, infrastructure development, pollution risks and overall water insecurity across the border. We observe that desalination – while seemingly attractive as a technical solution for increasing water supplies – fails to address water-security concerns in the region sustainably, or equitably. Instead, transnational disparities are emphasized, and manifest themselves in an uneven distribution of environmental externalities, such as degraded ecosystems and a steep rise in water prices. These would be shouldered by Mexican communities, as desalinated water is shipped across the border to the US. Such mismatches between who benefits from such water projects and who pays (via environmental impacts) may impede cooperation (Fischhendler & Feitelson, 2003).

As our Peruvian case illustrates, economic asymmetries and uneven political power, even within a single nation, can make equitable water security difficult to achieve. In the Ica River basin, while upstream communities might typically have greater control over water resources (Zeitoun & Warner, 2006), economic disparities are so dramatic they trump hydrologic position. Upstream communities, while closer to reliable water sources, are disadvantaged and experience water insecurity due to their low economic means. Downstream communities exert control via structural and financial means, building infrastructure to capture water from upstream basins. Amidst economic disparities, institutions that promote equitable allocation and benefit-sharing are critical. In multiple cases (e.g., the binational desalination, Catamayo-Chira and Andean glacier cases), the need for effective water-management institutions and policies – at subnational, national, and binational levels – was a key theme.

What have we learned about institutional responses to transboundary water-security challenges?

Overall, in our cases, we discern a broad range of institutional responses to transboundary water-security challenges, from informal cross-border cooperative scientific research, to social movements that inspire national legal frameworks, to binational frameworks being developed between historically contentious neighbours. Across all cases, the need for robust and well-coordinated institutions for water management was consistent. Addressing water-security challenges in particular calls for context-specific governance approaches (Gerlak et al., 2018) that are adaptive, multilevel and attentive to social power (Bakker & Morinville, 2013). Our case studies suggest that in transboundary contexts, institutional capacity is needed on multiple levels, and institutional arrangements and practices need to be flexible and responsive to changing or emerging conditions.

Effective institutional arrangements are often invoked to address transboundary water issues, such as conflict resolution, risk management and trade-off negotiation (Pahl-Wostl, Palmer, & Richards, 2013). Yet, institutional approaches are implemented at multiple levels. While transboundary water security can be addressed via collective action at the international level (Tarlock & Wouters, 2009), it can also be promoted by coordinating domestic laws and policies to address regional concerns (Mirumachi, 2013).

In the case of international neighbours, political relationships between national governments can evolve over time, but may be slow to transition from conflict to cooperation. Where international relationships remain strained, international organizations may play a crucial role to help facilitate cooperation as third parties. In the Upper Santa Cruz basin, in the absence of a legal framework for binational groundwater management, efforts to address transboundary groundwater have leveraged existing institutional capacity. The IBWC helped devise a framework for cooperation on binational aquifer assessment. The resultant TAAP engages binational academic institutions, government agencies and NGOs in cooperative scientific assessment. While imposing binational groundwater management is explicitly forbidden by the agreed framework (IBWC (International Boundary and Water Commission), 2009), scientific cooperation has other benefits: it can contribute to trust-building, participation, cross-border dialogue and relationship-building (Blomquist & Ingram, 2003; Gerlak & Mukhtarov, 2016).

In other instances, solutions to shared problems may exist at the local level. Locally based cooperation may be easier to implement and better reflect local needs (Eckstein, 2013). In the Ica River basin, we saw that subnational basins had better access to tools and mechanisms for addressing water security than transnational ones. However, local capacity has limitations. In the transboundary Catamayo-Chira basin, although land-use planning was conducted locally, its influence on transboundary water-management outcomes was limited without support from higher government levels. The design of binational cooperative frameworks has been a slow process, underlining the importance of institutional capacity on multiple levels. If shared interests of riparian nations, particularly those identified via local participation, can be leveraged into formal international cooperative processes, agreements and treaties may be more effective and resilient over time (Wolf, 2007).

Finally, in other cases still, local and regional social movements can influence national legislation, as demonstrated when Argentina passed a national glacier-conservation law, the first national glacier legislation of its kind. However, in this case, transboundary coordination is still lacking: thus far, Argentina and Chile have progressed differentially on glacier-conservation legislation. To fully address this transboundary challenge, coordinated binational efforts are needed.

Because our case studies all involve arid-to-semiarid environments, limited water supply is an important driver of water insecurity. While water scarcity can make it more difficult for nations to agree over shared resources (Hensel & Brochmann, 2007), it may also be a catalyst for cooperative efforts (Yoffe, Wolf, & Giordano, 2003) and new technological approaches. However, new technologies must be accompanied by appropriate institutional arrangements. In considering binational desalination in north-western Mexico, for example, we observe the need for institutions designed to address binational water allocation, payments, compensation, environmental impacts and trade to be in place at the start of such a project, accompanied by national-level policies for allocation and use of desalinated water. This illustrates how the viability of technological solutions depends on adequate institutional capacity at multiple levels – which may not exist initially – and stable binational political relations (Wilder et al., 2016).

Another key feature of institutional capacity is the ability of institutions to be resilient to change (Yoffe et al., 2003). In the transboundary context, investments in new capacity can create a feedback mechanism that helps reduce uncertainty and fosters further institutional adaptation (Heikkila, Gerlak, Bell, & Schmeier, 2013). Transboundary institutions that fail to adapt may ultimately fail to be effective (Berardo & Gerlak, 2012). Flexibility and adaptability by design can help address challenges effectively despite dynamic or changing conditions, such as those that result from climatic change.

Climate change is a key driver of water insecurity particularly in arid regions, creating new and evolving challenges that demand flexible, adaptive and responsive institutional arrangements (Eckstein, 2009). While surface water agreements, such as the 1944 Water Treaty, are designed with the flexibility to add incremental items (as with the 'minutes'), a binational agreement on shared groundwater management is unlikely to be similarly codified. In the absence of an adaptable legal mechanism for groundwater, binational institutions and networks, such as the TAAP and transboundary NGOs, can help guide groundwater management to be adaptable and responsive to climate-change impacts in the short term.

For high-elevation mountain water sources, climate-driven changes in the rate and timing of snow accumulation and glacier melt have had unpredictable effects, which may even result in redirection of flow among neighbouring basins (Shugar et al., 2017), possibly with transboundary impacts. In the Andean mountains, because glaciers span the border between Chile and Argentina, a coordinated policy response is needed at the international level. Until now, cross-border cooperation on glacier conservation has not been a priority, and national policies have developed slowly in this region. Sharing of scientific information and knowledge, which Argentina and Chile have already begun, is essential to fully understand the transboundary nature of glacier-stored water supplies and thereby address water security.

Adaptable institutional arrangements are also needed to redress inequity in water access (Pahl-Wostl et al., 2004). In the Ica River basin, water allocation practices are driven by economic and political power asymmetries. Infrastructure and water management policies are designed by and for powerful downstream communities. In the Catamayo-Chira River basin, access to potable water supplies is hindered by weak water provisioning and management. To improve water security, water-management policies must provide equitable benefits both upstream and downstream. Overall, while strategies involving enhanced institutional capacity and adaptive management involve high transaction costs associated with cooperation, dialogue and information-gathering, these may still be warranted when dealing with the complex issues characteristic of transboundary water (Raadgever, Mostert, Kranz, Interwies, & Timmerman, 2008).

Conclusions

Our close examination of five cross-border cases across the arid Americas reveals that each case features a distinct set of water-security attributes and corresponding context-specific challenges. Our analysis shows how the transboundary context, in a variety of configurations, can challenge planning, governance and implementation, while also permitting constructive, cooperative action across political barriers in some cases.

Many factors, individually and collectively, make addressing water security a complex endeavour in cross-border, or trans-jurisdictional, settings. The chief compounding factors related to the transboundary context are national-sovereignty concerns and uneven power dynamics; political, economic and physiographic asymmetries; and insufficient institutional capacity. We observe a broad range of institutional responses to these challenges – from informal cross-border scientific research, to social movements, to binational frameworks between historically contentious neighbours. Yet, we also see where institutional capacity is lacking, resulting in failures to promote international collective action, uneven benefit-sharing exacerbated by binational arrangements, and persistent inequity in water access.

Robust and well-coordinated institutional arrangements are needed to deal with the challenges of political contention, cross-border asymmetries and changing physical, social or institutional contexts, yet effective institutional arrangements will differ by location. Governance capacity at multiple levels can be relevant for transboundary water issues – from local cooperation to coordination of domestic policies and international collective action. In some cases, local, domestic or regional efforts can be effectively leveraged into cooperative agreements at the international level. In others, certain roles or functions (e.g., regulatory enforcement, data collection or participation) may be better suited for national to subnational levels and can address challenges when barriers inhibit collective action at the international level. Furthermore, institutions can be more effective and have greater longevity if they are flexible and adaptable to change.

However, building institutional capacity requires both resources and time. Other challenges may be that existing political frameworks and path-dependency may limit the options of institutional responses or that power asymmetries derived from differential economic development, financing, technology or resources may be difficult, if not impossible, to redress at the international level. Thus, it is imperative that the historical, cultural, political, economic and social aspects of the transboundary context are considered in determining feasible institutional responses in each context.

Our analysis is not without limitations. The water-security framing we employed enhanced our analysis, yet the water-security attributes listed in Table 1 are not comprehensive, nor do they represent a standardized list. Not all of the relevant attributes revealed by our case studies were represented – namely, cultural, social, institutional and information asymmetries, as well as power dynamics. We also recognize the limitations of our results due to our consideration of only five cases and our qualitative approach. Further research should examine other transboundary contexts in arid regions, and combine quantitative data with qualitative analysis. A useful next step in research would go beyond identifying the breadth and diversity of water-security attributes to examine the nature of their interactions and feedbacks via a system-dynamics approach (e.g., Kotir, Smith, Brown, Marshall, & Johnstone, 2016).

Nevertheless, our results can inform strategies for addressing transboundary water security in arid regions via robust, coordinated and adaptable institutional capacity at multiple levels, while demonstrating the distinctly context-specific nature of both water-security challenges. Distinguishing the many and varied components of complex transboundary water-security challenges is helpful to shed light on an equally broad range of context-appropriate – and ultimately effective – strategies for enhancing transboundary water security.

Acknowledgments

The maps in this manuscript were designed by Adriana Zuniga-Teran. The authors thank Elma Montaña for reviewing a portion of this manuscript. We also thank our anonymous reviewers for their insightful suggestions. This work was undertaken as part of the International Water Security Network, a project funded by Lloyd's Register Foundation, a charitable foundation helping protect life and property by supporting engineering-related education, public engagement and the application of research. We further acknowledge the Inter-American Institute for Global Change Research, for Project SGP-CRA005, supported by US National Science Foundation (NSF) Grant No. GEO-1138881. The paper also benefited from support from the Morris K. Udall and Stewart L. Udall Foundation, in Tucson, Arizona.

Notes

1. See Conca (2005), Young (1989), and Hayton and Utton (1989), whose seminal model groundwater treaty recognized what they called 'critical transboundary resource management areas'.
2. Among the earliest examples of such arrangements are the Colorado and Rio Grande (1944), Danube (1948), Indus (1960), Rhine (1963) and Senegal (1963) river basins, all of which created basin commissions or authorities, and some of which enacted treaties.
3. The case-study analysis relies on knowledge and experience developed through years of fieldwork and research projects conducted by the co-authors in these locations. Authors are active participants (or close affiliates) in two water-security networks operating in the Americas: the International Water Security Network, which is headquartered in Bristol, UK, and AQUASEC, the Inter-American Institute for Global Change Research centre of

excellence for water security in the Americas. Both were established in the early 2010s to study place-based water-security conditions and foster enhancements through science–policy dialogues.

4. Throughout this essay, we use the term 'arid' to refer generally to water-scarce areas. This includes regions that are semiarid (200–500 mm precipitation per year), to arid (25–200 mm), to hyperarid (less than 25 mm), according to the classic definition by Grove (1977).
5. These include the heated discussion over a border wall, immigration, narcotrafficking, and free trade – all of which have become more prominent since 2016.
6. The North American Free Trade Agreement went into effect in 1994.
7. Safe yield is defined in Arizona as 'a groundwater management goal which attempts to achieve and thereafter maintain a long-term balance between the annual amount of groundwater withdrawn in an active management area and the annual amount of natural and artificial recharge in the active management area' (A.R.S. §45–561).
8. The 1944 Water Treaty (Utilization of Waters of the Colorado and Tijuana Rivers and of the Rio Grande) allocated 1850 million cubic metres of the Colorado River's flow to Mexico, addressed allocations on the Rio Grande and Tijuana Rivers, and emphasized the need to prioritize transboundary sanitation issues. 'Minutes' are added to the treaty for incremental adjustments.
9. IBWC's Mexican analogue is CILA, the Comisión Internacional de Límites y Agua; thus, both countries have representatives in this binational institution.
10. The only formal agreement that addresses groundwater is IBWC's Minute 242, of 1973, which limits groundwater extraction in a small area near San Luis, AZ.
11. In 2006, the US Congress passed the Transboundary Aquifer Assessment Act (Public Law 109–448), authorizing the US Geological Survey, in partnership with federally recognized, public-university-based water-resources research centres in US border states, to cooperate with Mexico on this effort.
12. The first TAAP report, conducted on the San Pedro aquifer, the Binational Study of the Transboundary San Pedro Aquifer (Callegary et al., 2016), includes a binational database, technical analysis and binational maps, printed in both English and Spanish. The San Pedro aquifer is also shared between Arizona and Sonora, and is directly east of the Santa Cruz.
13. In the US and in Mexico, the largest subnational administrative units are called states; in Ecuador and Argentina, they are provinces; in Peru, they are departments. Chile has 15 large regions, which are divided into 54 provinces.
14. These may include urban expansion, changing climate and hydrologic regimes, invasive species, and shifting socio-political-economic processes.
15. Argentinean Institute of Nivology, Glaciology and Environmental Sciences.
16. The suit is highly ironic since it plays into the hands of the environmentalists' adversaries, the mining companies.

Disclosure statement

No potential conflict of interest was reported by the authors.

Funding

This work was supported by the Inter-American Institute for Global Change Research [SGP-CRA005];Lloyd's Register Foundation [CE-12-1051/CE-12-0801];National Science Foundation [GEO-1138881];Udall Foundation [MKU07066];

References

Abraham, E. M., & Villalba, R. (2008, October 2). Society, desertification and climate change in the Argentinean Chilean Central Andes Region. In *IHDP update, mountain regions: Laboratories for adaptation* (Issue 2, pp. 31–35). Bonn, Germany: Secretariat of the International Human Dimensions Programme on Global Environmental Change.

Aceituno, P. (1989). On the functioning of the Southern Oscillation in the South American sector. Part II. Upper-air circulation. *Journal of Climate, 2*(4), 341–355.

AECID (Agencia Española de Cooperación Internacional para el Desarrollo). (2005). *Proyecto Binacional Catamayo-Chira* Cuenca Binacional Catamayo-Chira [Catamayo-Chira Binational Basin]. Caracterización hídrica y adecuación entre la Oferta y la Demanda, Caracterización Territorial y Documentación Básica. Piura-Loja: Author..

Allan, J. A. (2002). *The Middle East water question: Hydropolitics and the global economy*. London, UK: I. B. Tauris.

Allan, J. A. (2011). *Virtual water: Tackling the threat to our planet's most precious resource*. London, UK: I. B. Tauris.

ANA (Autoridad Nacional del Agua). (2015). Informe técnico del cuarto monitoreo participativo de calidad de agua superficial de la cuenca Chira [Technical report of the fourth participatory monitoring on water quality in surface water of the Chira Basin]. In *Informe técnico*. Piura: Author.

Bakker, K. (2012). Water security: Research challenges and opportunities. *Science, 337*(6097), 914–915.

Bakker, K., & Morinville, C. (2013). The governance dimensions of water security: A review. *Philosophical Transactions of the Royal Society A: Mathematical, Physical and Engineering Sciences, 371*(2002), 20130116.

Bates, B.C., Kundzewicz, Z.W., Wu, S., & Palutikof, J. P. (Eds.). (2008). Climate change and water. *Technical paper of the intergovernmental panel on climate change*. Geneva: IPCC Secretariat.

Berardo, R., & Gerlak, A. K. (2012). Conflict and cooperation along international rivers: Crafting a model of institutional effectiveness. *Global Environmental Politics, 12*(1), 101–120.

Binational Development Plan. (2016). *Plan Binacional de Desarrollo de la Region Fronteriza Perú-Ecuador* [Binational Development Plan for the Peru-Ecuador Border Region]. Capítulo Perú. Retrieved from http://planbinacional.org.pe/plan-binacional-2/marco-juridico/

Blatter, J., & Ingram, H. (Eds.). (2001). *Reflections on water: New approaches to transboundary conflicts and cooperation* (pp. 356). Cambridge: MIT Press.

Blomquist, W., & Ingram, H. M. (2003). Boundaries seen and unseen: Resolving transboundary groundwater problems. *Water International, 28*(2), 162–169.

Boelens, R., & Vos, J. (2012). The danger of naturalizing water policy concepts: Water productivity and efficiency discourses from field irrigation to virtual water trade. *Agricultural Water Management, 108*, 16–26.

Callegary, J., Minjárez Sosa, I., Tapia Villaseñor, E., dos Santos, P., Monreal Saavedra, R., Grijalva Noriega, F., … Leenhouts, J. (2016). Binational study of the transboundary San Pedro aquifer. *International Boundary and Water Commission*, p. 170. Retrieved from http://ibwc.gov/Files/San_Pedro_River_Binational%20Report_013116.pdf

Cárdenas Panduro, A. I. (2012). *'La Carrera Hacia el Fondo': Acumulación de agua subterranea por empresas agro-exportadoras en el Valle de Ica, Perú* [Groundwater accumulation by agro-export businesses in the Ica Valley, Peru] (Master's thesis). Irrigation and Water Engineering, International Land and Water Management, Wageningen University, The Netherlands.

CONAGUA. (2016). *Situación del Subsector Agua Potable, Drenaje y Saneamiento, edición 2016* [State of the potable water, drainage and sanitation sub-sector]. Mexico: Author.

Conca, K. (2005). *Governing water: Contentious transitional politics and global institution building*. Cambridge, MA: MIT Press.

Cook, C., & Bakker, K. (2012). Water security: Debating an emerging paradigm. *Global Environmental Change, 22*(1), 94–102.

Damonte, G., Pacheco, E., & Grados, C. (2014). Dinámicas de concentración y escasez de agua: El boom agroexportador y los pequeños propietarios en las zonas media y alta del Río Ica [Water concentration and water shortage dynamics: The agro-export boom and the small landholders in the medium and upper zones of the Ica River]. In M. T. Oré & G. Damonte (Eds.), *¿Escasez del Agua? Retos Para la Gestión de la Cuenca del Río Ica* (pp. 127–172). Lima: Fondo Editorial de la Pontificia Universidad Católica del Perú.

Damonte Valencia, G. H. (2015). Redefinition of hydro-social territories: Hydric Control in the Ica Valley, Peru (1993–2013). *Cuadernos de Desarrollo Rural, 12*(76), 109–133.

Dibble, S. (2014). One desal plant, two countries? Private proposal for a plant in Rosarito Beach moves forward. *San Diego Union Tribune*, August 24.

Eckstein, G. E. (2009). Water scarcity, conflict, and security in a climate change world: Challenges and opportunities for international law and policy. *Wisconsin International Law Journal, 27*, 409.

Eckstein, G. E. (2013). Rethinking transboundary ground water resources management: A local approach along the Mexico-US border. *Georgetown International Environmental Law Review, 25*(1), 95–128.

EPA & SEMARNAT. (2013). *Programa Ambiental México- Estados Unidos: Frontera 2020* [Environmental Porgram Mexico-United States: Border 2020]. Tijuana: EPA- SEMARNAT.

Falkenmark, M., & Lundqvist, J. (1999). Towards upstream/downstream hydrosolidarity. *Proceedings of the 1999 SIWI Seminar*. Stockholm, Sweden: SIWI.

Feitelson, E., & Haddad, M. (Eds.). (2001). *Management of shared groundwater resources: The Israeli-Palestinian case with an international perspective.* Boston, MA: Kluwer Academic.

Fischhendler, I., & Feitelson, E. (2003). Spatial adjustment as a mechanism for resolving river basin conflicts: The US–Mexico case. *Political Geography, 22*(5), 557–583.

Food and Agriculture Organization (FAO). (1996, November 13–17). *World food summit plan of action*. Rome, Italy: World Food Summit.

Gain, A. K., Giupponi, C., & Wada, Y. (2016). Measuring global water security towards sustainable development goals. *Environmental Research Letters*, 11(12), 124015.

Garfin, G., Franco, G., Blanco, H., Comrie, A., Gonzalez, P., Piechota, T., … Waskon, R. (2014). Ch. 20: Southwest. In J. M. Melillo, T. Richmond, & G. W. Yohe (Eds.), *Climate change impacts in the United States: The third national climate assessment* (pp. 462–486). Washington, DC: US Global Change Research Program.

Geng, D. (2016). *Reformas institucionales del agua en la costa peruana: Análisis de la gobernanza del agua en la cuenca Ica-Alto Pampas* [Institutional reforms on water in the Peruvian coast: Governance analysis in the Ica-Alto Pampas Basin] (Masters thesis). Water Resources Management, Pontificia Universidad Católica del Perú.

Gerlak, A. K. (2004). One Basin at a Time: The global environment facility and governance of transboundarywaters. *Global Environmental Politics, 4*(4), 108–141.

Gerlak, A. K., House-Peters, L., Varady, R. G., Albrecht, T., Zúñiga-Terán, A., de Grenade, R. R., … Scott, C. A. (2018). Water security: A review of place-based research. *Environmental Science and Policy, 82*(February), 79–89.

Gerlak, A. K., & Mukhtarov, F. (2016). Many faces of security: discursive framing in cross-border natural resource governance in the Mekong River Commission. *Globalizations, 13*(6), 719–740. doi:10.1080/14747731.2015.1134133

Gleick, P. (2015). The future of desalination in California is still in the future: California, Israel, and Australia. *National Geographic Science Blogs*. Pacific Institute. Retrieved from: http://pacinst.org/national-geographic-scienceblog-the-future-of-desalination-in-california-is-still-in-the-future-california-israel-and-australia/

Global Water Partnership (GWP). (2000). *Towards water security: A framework for action.* Stockholm: GWP Secretariat.

Grey, D., & Sadoff, C. W. (2007). Sink or swim? Water security for growth and development. *Water Policy, 9*(6), 545–571.

Grove, A. T. (1977). Desertification. *Progress in Physical Geography, 1*(2), 296–310.

Hamner, J. H., & Wolf, A. T. (1998). Patterns in international water resource treaties: The transboundary freshwater dispute database. *Colorado Journal of International Environmental Law and Policy*, *9*(Suppl.), 157–177.

Hayton, R. D., & Utton, A. E. (1989). Transboundary groundwaters: The Bellagio Draft Treaty. *Natural Resources Journal*, *3*, 663–722.

Heikkila, T., Gerlak, A. K., Bell, A., & Schmeier, S. (2013). Adaptation in a transboundary river basin: Linking stressors and adaptive capacity within the Mekong River Commission. *Environmental Science and Policy*, *25*, 73–82.

Hensel, P. R., & Brochmann, M. (2007, March). *Armed conflict over international rivers: The onset and militarization of river claims*. Paper presented at the annual meeting of the International Studies Association, Chicago, IL.

Hepworth, N. D., Postigo, J. C., Güemes Delgado, B., & Kjell, P. (2010). *Drop by drop: Understanding the impacts of the UK's water footprint through a case study of peruvian asparagus*. London: Progressio, Centro Peruano de Estudios Sociales (CEPES) and Water Witness International.

Herz Sáenz, C., & Huamán Arias, A. (2014). *Perú/Ica y Huancavelica: Estudio de Percepciones Sobre la Problemática del Agua en la Cuenca del Río Tambo-Santiago-Ica* [Peru/Ica and Huancavelica: A study on the perceptions of water challenges in the Tambo-Santiago-Ica River Basin]. Proyecto DIRMAPA: Diálogo Regional sobre Manejo Ambiental y de Recursos Naturales en los Países Andinos. Lima: Cooperación Alemana al Desarrollo (GIZ).

Hirt, P., Snyder, R., Hester, C., & Larson, K. (2017). Water consumption and sustainability in Arizona: A tale of two desert cities. *Journal of the Southwest*, *59*(1–2), 264–301.

Hocquenghem, A., & Durt, E. (2002). Integración y desarrollo de la región fronteriza peruano ecuatoriana: Entre el discurso y la realidad, una visión local [Integration and development of the transboundary region of Ecuador and Peru: Between the discourse and reality, a local vision]. *Revista del Instituto Francés de Estudios Andinos*, *31*(1), 39–99.

Hurlbert, M. A. (2018). *Adaptive governance of disaster: Drought and flood in rural areas*. Cham: Springer.

INEGI (Instituto Nacional de Estadística y Geografía). (2010). *Panorama Sociodemográfico Sonora* [Sociodemographic overview of Sonora]. Retrieved April 16, 2017, from http://www3.inegi.org.mx/sistemas/mexicocifras/default.aspx?e=26

INEI (Instituto Nacional de Estadística e Informática). (2007). *Población censada, según departamento y año censal 2007* [Census population, according to the department and census year 2007]. Retrieved June 22, 2017, from http://www.inei.gob.pe/estadisticas/indice-tematico/poblacion-y-vivienda/

Ingram, H. M., Laney, N. R., & Gilliland, D. M. (1995). *Divided waters: Bridging the US-Mexico border*. Tucson: University of Arizona Press.

International Boundary and Water Commission. (2009, August 19). *Joint report of the principal engineers regarding the joint cooperative process United States-Mexico for the transboundary aquifer assessment program: 11p.*

International Boundary and Water Commission. (2012). *Minute 391: Interim international cooperative measures in the Colorado River Basin through 2017 and extension of minute 318 cooperative measures to address the continued effects of the April 2010 Earthquake in the Mexicali Valley, Baja California*. International Boundary and Water Commission United States and Mexico. Retrieved from https://www.ibwc.gov/Files/Minutes/Min319_Env_Fact_Sheet.pdf

IPCC (Intergovernmental Panel on Climate Change). (2007). *Synthesis report, contribution of Working Groups I, II and III to the fourth assessment report of the Intergovernmental Panel on Climate Change*. [Core Writing Team, Pachauri RK, Reisinger A (eds)] (p. 104). Geneva, Switzerland: IPCC.

IPCC (Intergovernmental Panel on Climate Change). (2014). *Climate change 2014: Synthesis report. contribution of working Groups I, II and III to the fifth assessment report of the intergovernmental panel on climate change*. [Core Writing Team, R.K. Pachauri and L.A. Meyer, L. (eds.)] (p. 151). Geneva, Switzerland: IPCC.

Isla Raffaele, M. L. (2016). El proceso de juridificación de la Ley de Glaciares en las disputas por la Megaminería en Argentina [The process of juridification of the Glaciers Law in the disputes for Megamining in Argentina]. *Ecología Política*, 103–107. Retrieved from http://www.ecologiapolitica.info/?p=6067

Jepson, W. (2014). Measuring "no-win" waterscapes: Experience-based scales and classification approaches to assess household water security in colonias on the US–Mexico border. *Geoforum*, *51*, 107–120.

Kapoor, M. (2017, March 30). *Nogales has a sewage problem*. Retrieved May 28, 2018, from https://www.hcn.org/articles/nogales-has-a-sewage-problem

Kelly-Richards, S. H., & Banister, J. M. (2017). A state of suspended animation: Urban sanitation and water access in Nogales, Sonora. *Political Geography*, *58*, 104–113.

Kennedy, J., Rodríguez-Burgueño, J. E., & Ramírez-Hernández, J. (2017). Groundwater response to the 2014 pulse flow in the Colorado River Delta. *Ecological Engineering*, *106*, 715–724.

Khadim, A. N. (2016). Defending Glaciers in Argentina. *Peace Review*, *28*(1), 65–75.

Kliot, N., Shmueli, D., & Shamir, U. (2001). Institutions for management of transboundary water resources: Their nature, characteristics and shortcomings. *Water Policy*, *3*, 229–255.

Kotir, J. H., Smith, C., Brown, G., Marshall, N., & Johnstone, R. (2016). A system dynamics simulation model for sustainable water resources management and agricultural development in the Volta River Basin, Ghana. *Science of the Total Environment*, *573*, 444–457.

Lautze, J. and Manthrithilake, H. (2012). Water security: Old concepts, new package, what value? *Natural Resources Forum* 36(2), 76-87.

Law 26.639. (2010). *Congress of the Nation, Argentina*. Retrieved from http://servicios.infoleg.gob.ar/infolegInternet/anexos/170000-174999/174117/norma.htm

Law 7722. (2007). *Province of Mendoza, Argentina*. Retrieved from http://www.legislaturamendoza.gov.ar/busqueda/_lib/tmp/sc_pdf_20170727134822_122_pdfreport_ley_sin_editar.pdf

Leichenko, R. M., & O'Brien, K. L. (2008). *Environmental change and globalization- double exposures*. New York: Oxford University Press.

Linton, J., & Brooks, D. (2011). Governance of transboundary aquifers: New challenges and new opportunities. *Water International*, *36*(March 2015), 606–618.

Maestre, F. T., Salguero-Gomez, R., & Quero, J. L. (2012). It is getting hotter in here: Determining and projecting the impacts of global environmental change on drylands. *Philosophical Transactions of the Royal Society B: Biological Sciences*, *367*(1606), 3062–3075.

Marty, F. (2001). *Managing international rivers: Problems, politics and institutions*. Bern: Peter Lang.

Masiokas, M. H., Villalba, R., Luckman, B. H., Le Quesne, C., & Aravena, J. C. (2006). Snowpack variations in the central Andes of Argentina and Chile, 1951–2005: Large-scale atmospheric influences and implications for water resources in the region. *Journal of Climate*, *19*(24), 6334–6352.

McEvoy, J. (2014). Desalination and water security: The promise and perils of a technological fix to the water crisis in Baja California Sur, Mexico. *Water Alternatives*, *7*(3), 518–541.

McEvoy, J. (2015). Can the adoption of desalination technology lead to aquifer preservation? A case study of a sociotechnical water system in Baja California Sur, Mexico. *Water*, *7*(10), 5224–5238.

Meerganz von Medeazza, G. L. (2005). Direct and socially-induced environmental impacts of desalination. *Desalination*, *185*(1–3), 57–70.

Megdal, S. B., & Petersen-Perlman, J. (2018, August). Groundwater governance and assessment in a transboundary setting. In V. I. Grover & G. Krantzberg (Eds.), *Lake governance* (pp. 48–72). The Netherlands: CRC/Taylor and Francis.

Megdal, S. B., & Scott, C. A. (2011). The importance of institutional asymmetries to the development of binational aquifer assessment programs: The Arizona-Sonora experience. *Water (Switzerland)*, *3*(3), 949–963.

Meixner, T., Manning, A. H., Stonestrom, D. A., Allen, D. M., Ajami, H., Blasch, K. W., … Walvoord, M. A. (2016). Implications of projected climate change for groundwater recharge in the western United States. *Journal of Hydrology*, *534*, 124–138.

Meza, F. J., Wilks, D. S., Gurovich, L., & Bambach, N. (2012). Impacts of climate change on irrigated agriculture in the Maipo Basin. *Chile: Reliability of Water Rights and Changes in the Demand for Irrigation. Journal of Water Resources Planning and Management, 138*(5), 421–430.

Milich, L., & Varady, R. G. (1999). Openness, sustainability, and public participation: New designs for transboundary river basin institutions. *The Journal of Environment & Development, 8*(3), 258–306.

Milman, A., & Ray, I. (2011). Interpreting the unknown: Uncertainty and the management of transboundary groundwater. *Water International, 36*(5), 631–645.

Milman, A., & Scott, C. A. (2010). Beneath the surface: Intranational institutions and management of the United States - Mexico transboundary Santa Cruz aquifer. *Environment and Planning C: Government and Policy, 28*(3), 528–551.

Mirumachi, N. (2013). Transboundary water security: Reviewing the importance of national regulatory and accountability capacities in international transboundary river basins. In B. Lankford, K. Bakker, M. Zeitoun, & D. Conway (Eds.), *Water security: Principles, perspectives, and practices* (pp. 166–180). New York, NY: Routledge.

Morehouse, B. J. (1995). A functional approach to boundaries in the context of environmental issues. *Journal of Borderlands Studies, 10*(2), 53–73.

Mumme, S. P. (2015). The evolution of natural resource conservation capacity on the U.S.-Mexico border: Bilateral and trilateral environmental agreements since La Paz. *Review of Policy Research, 32*(1), 19–39.

Nelson, K. (2007, February). *Groundwater flow model of the Santa Cruz active management area along the effluent-dominated Santa Cruz River, Santa Cruz and Pima Counties, Arizona* (Modeling Report No. 14). Arizona Department of Water Resources. Retrieved from https://repository.asu.edu/attachments/92801/content/SantaCruzAMA_GroundwaterFlowModel.pdf

Norman, E., Bakker, K. J., Cook, C., Dunn, G., & Allen, D. (2010). *Water security: A primer.* Program on Water Governance. Vancouver, CA: University of British Columbia.

Ocasiot, K. M. O. (2015). Feeling salty? Regulating desalination plants in the United States and Spain. *Cornell International Law Journal, 48(2)*, 451–479.

OECD. (2013). *Water security for better lives: A summary for policymakers.* Paris: OECD Publishing.

Pahl-Wostl, C., Bouwen, R., Craps, M., Morel, P., Mostert, E., Ridder, D., & Taillieu, T. (2004). The importance of processes of social learning for transboundary water management. Perspectives from the HarmoniCOP project. In *Proceedings from the Conference on integrated water management of transboundary catchments; TRANSCAT: Venice, Italy.*

Pahl-Wostl, C., Palmer, M., & Richards, K. (2013). Enhancing water security for the benefits of humans and nature-the role of governance. *Current Opinion in Environmental Sustainability, 5*(6), 676–684.

Palma, A., González, F., & Cruickshank, C. (2015). Managed aquifer recharge as a key element in Sonora River Basin management, Mexico. *Journal of Hydrologic Engineering, 20*(3), B4014004.

PETACC (Proyecto Especial Tambo-Ccaracocha). (2010). *Estudio de Factibilidad del Proyecto de Inversion Pública 'Construcción de la Presa Tambo'* [A feasibility study on the public investment project 'Building of the Tambo Dam']. Volumen 1: Resumen Ejecutivo. Ica, Peru: Regional Government of Ica.

PNUD (Programa de las Naciones Unidas para el Desarrollo; or UNDP). (2015). *Proyecto "Gestión Integrada de Recursos Hídricos de las Cuencas Transfronterizas y Acuíferos de Puyango-Tumbes, Catamayo-Chira y Zarumilla"* [The 'integrated water resources management in transboundary Basins and aquifers of Puyango-Tumbes, Catamayo-Chira and Zarumilla' project] (Documento de Proyecto). Quito: PNUD.

Pombo, A., Breceda, A., & Valdez Aragón, A. (2008). Desalinization and wastewater reuse as technological alternatives in an arid, tourism booming region of Mexico. *Frontera Norte, 20* (39), 191–216.

Raadgever, G. T., Mostert, E., Kranz, N., Interwies, E., & Timmerman, J. G. (2008). Assessing management regimes in transboundary river basins: Do they support adaptive management?

Ecology and Society: A Journal of Integrative Science for Resilience and Sustainability IF - 2.855, 13(1), 1–21. Retrieved from http://www.ecologyandsociety.org/vol13/iss1/art14/

Richardson, K., Steffen, W., & Liverman, D. (Eds.). (2011). *Climate change: Global risks, challenges and decisions*. Cambridge: Cambridge University Press. doi:10.1017/CBO9780511973444

Sadoff, C., & Grey, D. (2002). Beyond the river: The benefits of cooperation on international rivers. *Water Policy, 4*, 389–403.

Sanchez, R., & Eckstein, G. (2017). Aquifers shared between Mexico and the United States: Management perspectives and their transboundary nature. *Groundwater*, 1–11. doi:10.1111/gwat.12533

Schmeier, S. (2013). *Governing international watercourses: River Basin organizations and the sustainable governance of internationally Shared Rivers and Lakes*. London, UK: Routledge.

Scott, C., Megdal, S., Oroz, L., Callegary, J., & Vandervoet, P. (2012a). Effects of climate change and population growth on the transboundary Santa Cruz aquifer. *Climate Research, 51*(2), 159–170.

Scott, C. A., & Banister, J. M. (2008). The dilemma of water management 'regionalization' in Mexico under centralized resource allocation. *International Journal of Water Resources Development, 24*(1), 61–74.

Scott, C. A., Meza, F. J., Varady, R. G., Tiessen, H., McEvoy, J., Garfin, G. M., … Montaña, E. (2013). Water security and adaptive management in the arid Americas. *Annals of the Association of American Geographers, 103*(2), 280–289.

Scott, C. A., Varady, R. G., Meza, F., Montaña, E., de Raga, G. B., Luckman, B., & Martius, C. (2012b, May–June). Science-policy dialogues for water security: Addressing vulnerability and adaptation to global change in the Arid Americas. *Environment: Science and Policy for Sustainable Devel, 54*(3), 30–42.

Shamir, E., Megdal, S. B., Carrillo, C., Castro, C. L., Chang, H. I., Chief, K., … Prietto, J. (2015). Climate change and water resources management in the Upper Santa Cruz River, Arizona. *Journal of Hydrology, 521*, 18–33.

Shugar, D. H., Clague, J. J., Best, J. L., Schoof, C., Willis, M. J., Copland, L., & Roe, G. H. (2017). River piracy and drainage basin reorganization led by climate-driven glacier retreat. *Nature Geoscience, 10*(5), 370–375.

Surfleet, C. G., & Tullos, D. (2013). Variability in effect of climate change on rain-on- snow peak flow events in a temperate climate. *Journal of Hydrology, 479*, 24–34.

Talliant, J. D. (2013). The human right… to glaciers? *Journal of Environmental Law and Litigation, 28*(1), 59–78.

Tarlock, D., & Wouters, P. (2009). Reframing the water security dialogue. *Journal of Water Law, 20*(2–3), 53–60.

Toledo, A. (2017, July). Experiencias de desarrollo institutcional en la gestion de aguas transfronterizas: Experiencia Peruana [Experiences of institutional development in the management of transboundary waters: The Peruvian Experience]. Presented at Quintas Jornadas de Derecho de Aguas, Lima, Peru. Retrieved from http://jornada.pucp.edu.pe/derecho-de-aguas/wp-content/uploads/sites/8/2013/07/1.-Adolfo-Toledo.pdf

Udall, S. L., & Varady, R. G. (1994). Environmental conflict and the world's new international borders. *Transboundary Resources Report, 7*(3), 5–6.

Uitto, J. I., & Duda, A. M. (2002). Management of transboundary water resources: Lessons from international cooperation for conflict prevention. *Geographical Journal, 168*(3), 365–378.

United Nations Educational, Scientific and Cultural Organization (UNESCO). (2013). *Free flow: Reaching water security through cooperation*. Paris: UNESCO and Tudor Rose.

United Nations Environment Program (UNEP). (2009). *Water security and ecosystem services: The critical connection*. Nairobi: UNEP.

UN-Water. (2013). *Water security and the global water agenda: A UN-water analytical brief.* Hamilton, ON: UN University.

Urkidi, L. (2010). A glocal environmental movement against gold mining: Pascua–Lama in Chile. *Ecological Economics, 70*(2), 219–227.

Urteaga, P. (2014). Creadores de paisajes hídricos: Abundancia de agua, discursos y mercado en las cuencas de Ica y Pampas [Creators of water landscapes: Water abundance, discourses and markets in the Ica and Pampas basins]. In M. T. Oré & G. Damonte (Eds.), *¿Escasez del Agua? Retos Para la Gestión de la Cuenca del Río Ica* (pp. 227–267). Lima: Fondo Editorial de la Pontificia Universidad Católica del Perú.

US Census. (2015). *QuickFacts Nogales city, Arizona*. Retrieved April 29, 2017, from http://www.census.gov/quickfacts/table/PST045215/0449640

Van der Zaag, P., & Savenije, H. H. G. (2000). Towards improved management of shared river basins: Lessons from the Maseru conference. *Water Policy*, *2*, 47–63.

Varady, R., & Ward, E. (2009). Transboundary conservation in the borderlands: What drives environmental change. In L. Lopez-Hoffman, E. D. McGovern, R. G. Varady, & K. W. Flessa (Eds.), *Conservation of shared environments: Learning from the United States and Mexico* (pp. 9–22). Tucson: University of Arizona Press.

Varady, R. G., & Morehouse, B. J. (2003). Moving borders from the periphery to the center: river basins, political boundaries, and water management policy. In R. Lawford, D. Fort, H. Hartmann, & S. Eden (Eds.), *Water: Science, Policy, and Management* (pp. 143–159). Washington, DC: American Geophysical Union.

Varady, R. G., & White, D. R. (1992). The Arizona-Sonora border: Context and background. In R. G. Varady (Ed.), *Preserving Arizona's environmental Heritage, 59th Arizona Town hall research report* (Chapter 12, pp. 201–206). Tucson: University of Arizona Press.

Varis, O., Keskinen, M., & Kummu, M. (2017). Four dimensions of water security with a case of the indirect role of water in global food security. *Water Security*, *1*, 36–45.

Wagner, L. (2017). Megaminería y conflictos socioambientales en Argentina: Por el agua y más allá [Megamining and scoio-environmental conflicts in Argentina; for water and beyond]. In L. Bottaro & M. Sola Álvarez (coords.), *Agua y Megaproyectos mineros en América Latina* [Water and mining megaprojects in Latin America] (pp. 64–89). Buenos Aires, Argentina: Universidad Nacional de General Sarmiento.

Wheater, H. S., & Gober, P. (2015). Water security and the science agenda. *Water Resources Research*, 51(7), 5406–5424.

Wilder, M. O., Aguilar-Barajas, I., Pineda-Pablos, N., Varady, R. G., Megdal, S. B., McEvoy, J., … Scott, C. A. (2016). Desalination and water security in the US–Mexico border region: Assessing the social, environmental and political impacts. *Water International*, *41*(5), 756–775.

Wilder, M. O., Slack, J., Varady, R. G., Scott, C. A., Prichard, A., Morehouse, B., … Beaty, R. (2012). Urban water vulnerability and institutional challenges in Ambos Nogales. In M. O. Wilder, C. A. Scott, N. Pineda- Pablos, R. G. Varady, & G. M. Garfin (Eds.), *Moving forward from vulnerability to adaptation: Climate change, drought, and water demand in the urbanizing Southwest United States and Northern Mexico* (pp. 19–53). Tucson, AZ: Udall Center for Studies in Public Policy.

Wolf, A. T. (1998). Conflict and cooperation along international waterways. *Water Policy*, *1*(2), 251–265.

Wolf, A. T. (2007). Shared waters: Conflict and cooperation. *Annual Review of Environment and Resources*, *32*(2), 3.1–3.29.

Wolf, A. T., Natharius, J. A., Danielson, J. J., Ward, B. S., & Pender, J. K. (1999). International river basins of the world. *International Journal of Water Resources Development*, *15*(4), 387–427.

Yin, R. K. (2003). *Case study research: Design and methods* (3rd ed.). Thousand Oaks, CA: Sage publications.

Yin, R. K. (2017). *Case study research and applications: Design and methods* (6th ed.). Thousand Oaks, CA: Sage publications.

Yoffe, S., Wolf, A. T., & Giordano, M. (2003). Conflict and cooperation over international freshwater resources: Indicators of basins at RISR. *JAWRA Journal of the American Water Resources Association*, *39*(5), 1109–1126.

Young, O. R. (1989). *International cooperation: Building regimes for natural resources and the environment* (Cornell Studies in Political Economy). Ithaca, NY: Cornell University Press.

Zeitoun, M. (2011). The global web of national water security. *Global Policy*, 2(3), 286-296.

Zeitoun, M., Lankford, B., Krueger, T., Forsyth, T., Carter, R., Hoekstra, A. Y., … Matthews, N. (2016). Reductionist and integrative research approaches to complex water security policy challenges. *Global Environmental Change, 39*(July), 143–154.

Zeitoun, M., & Mirumachi, N. (2008). Transboundary water interaction I: Reconsidering conflict and cooperation. *International Environmental Agreements: Politics, Law and Economics, 8*(4), 297–316.

Zeitoun, M., & Warner, J. (2006). Hydro-hegemony - A framework for analysis of transboundary water conflicts. *Water Policy, 8*(5), 435–460.

De-nationalization and de-securitization of transboundary water resources: the Israeli–Palestinian case

David B. Brooks and Julie Trottier

Many people think of transboundary water in terms of national security. However, water is not, nor is it likely to become, a cause of war. Rather, the need is for water security, which implies that water management must balance the goals of efficiency, equity, sustainability and implementability. This article suggests how a joint management structure for fresh water can be designed to promote ongoing resolution of issues, and do so in a way that de-nationalizes and de-securitizes transboundary water. Though designed with the Israeli–Palestinian case in mind, the approach is applicable wherever water divides rather than unites states or peoples.

> New forms of hydro-diplomacy are desperately needed in order to address the lack of political will, financial resources and effective governance.
>
> Axworthy and Sandford (2012, p. 4)

The notion of securitizing water resources has a number of possible meanings, some of which are more, and some less, useful to water discourse and thus to transboundary water management. This paper will begin by reviewing three possible meanings for the concept of securitization and evaluating their relevance to water management. Second, it will consider how common meanings of securitization as well as fixed quantitative divisions of water resources subvert more efficient, equitable and sustainable use of shared water, as well as running afoul of established, local forms of water management. The remainder of the paper focuses on Israel and Palestine to illustrate that ways do exist to avoid these problems. (Here "Palestine" is used to mean both the West Bank and Gaza Strip as they currently exist, and the future State of Palestine within national borders that, with some adjustments, are likely to be similar to the armistice lines before the 1967 War.) The third section of the paper describes an institutional structure that is designed to avoid the defects of securitization and fixed allocations. Though this section incorporates conditions specific to Israel and Palestine, this focus is not intended to restrict more general applicability of either the analysis or the results. The final section goes on to suggest how to begin implementation of such a structure. An appendix provides highly summarized information on the water resources shared by Israel and Palestine.

The purpose of this paper is to suggest new concepts to explore when transboundary fresh water cannot simply be divided by convenient geographic features. However, because application of these concepts must be sensitive to local management, regional hydrogeology, and national power structures, it is not intended to urge action outside the area of the case study.

Full information on the proposed agreement on fresh water between Israel and Palestine has been published by Friends of the Earth Middle East, one of the few remaining Israeli-Palestinian-Jordanian organizations (Brooks & Trottier, 2012). A summary version has also been published (Brooks, Trottier, & Doliner, 2013) and is available for free download. http://www.tandfonline.com/doi/full/10.1080/02508060.2013.810038

Three meanings of "securitization"

When applied to water, the concept of securitization can have three different meanings.

- In its most common sense, it refers to a causal link between the availability of water resources in some nation and that nation's military or economic strength.
- In a more limited sense, it refers to competition among different stakeholders (or economic sectors, or regions, or communities) within a nation for assured quantities of fresh water for members of the relevant group.
- In a third sense, it is synonymous with "water security", a broad concept that implies adequate supplies of water of appropriate quality for household and commercial uses and for the protection of the ecosystems within which those activities occur.

The relevance of the first of the three meanings is questionable. History demonstrates that riparian states around the world prefer to cooperate over transboundary water bodies rather than to fight over them (Gleick, 2000; Jägerskog, 2003; Katz, 2011; Kliot, Shmueli, & Shamir, 2001; Wolf, 1998, 2007). Perhaps "cooperate" is too benevolent a word. In Asia, political power allows stronger nations (China in East Asia; India in South Asia; Kazakhstan and Uzbekistan in Central Asia) to take pretty much any water they want (Chellaney, 2011). The situation on the ground tends to some sort of equilibrium, but not to war. Further, as nation-states get used to joint water management, the process itself seems to have "a singularly important role to play both in facilitating the rebuilding of trust following conflict and in preventing a return to conflict" (Weinthal, Troell, & Nakayama, 2011, p. 149). Finally, careful analysis shows that per capita availability of fresh water is not particularly useful as an indicator either of national well-being or of national wealth (Chenoweth, 2008).

The situation among the Jordan Valley countries is not much different from what is found elsewhere in the world. Wolf (1998) reviewed a number of recent studies that purported to find a causal link between war and water, and he commented (p. 253):

> The examples most widely cited are wars between Israel and her neighbors.... The only problem with these theories is a complete lack of evidence.

In support of Wolf's argument, Fischhendler, Wolf, and Eckstein (2012) showed that Israeli–Arab water issues have been resolved by agreements that use imaginative language to dodge the most contentious aspects. Haddadin (2011) emphasized that Annex 2 in the Israel–Jordan Peace Treaty,[1] which focuses on water, not only facilitated agreement on other terms of the treaty but continues to be honoured even when other parts are not.

The absence of water wars does not mean that fresh water is free of conflict. Far from it! However, the conflict is mainly *intra*national rather than *inter*national, which is the second meaning for securitization. Cities that want water for domestic use can find themselves in conflict with farmers who want water for irrigation. Farmers who line their earthen irrigation canals with cement can prevent leakage but also block infiltration of water that has been flowing through an aquifer to the wells of a nearby village. Environmentalists who argue for leaving more water *in situ* to ensure continued delivery of environmental services will commonly be opposed by those who want the water for commercial use. These kinds of conflicts occur every day across the world. They need to be resolved, but the resulting issues regarding allocation of water are not essentially different from those regarding the location of new highways and other land-use choices. Key issues include comparison of benefits and costs, and impacts related to scale, to class and to gender – and of course to protection of the environment. They are profoundly political issues, and they are fraught with conflict. As Axworthy and Sandford concluded (2012, p. 12):

> The fact that there has been widespread international cooperation over water should not allow policy-makers to underestimate the complexity of the relationship between water and national security.

The same idea is contained in the adage commonly but incorrectly attributed to Mark Twain, "*Whiskey's for drinking; water's for fighting over*" (http://nsla.nevadaculture.org/index.php?option=com_content&task = view&id = 803&Itemid = 418). Complex they may be, and on occasion they may have led to fighting, but these sorts of conflicts have little or no relevance for national military or economic strength.

This article will adhere most closely to the third and arguably most useful meaning of water securitization – when it becomes equivalent to water security. Water security has both quantitative and qualitative dimensions, and is relevant to all uses of water, and is the only meaning of "securitization" that fully incorporates environmental services derived from water *in situ*. Karen Bakker (2012) recently explored recent use of the concept of water security in a short article that focused on research challenges and opportunities. She found that current work on water security is hindered by: (1) "multiple, and at times incommensurate, definitions of water security"; (2) "socio-environmental" issues that require "interdisciplinary, collaborative research" to provide "science-informed policy"; and (3) the need to bridge the "scalar mismatch" between hydrological focus on watersheds and political focus on the nation-state (Bakker, 2012, pp. 914, 915).

No doubt, until more of the research gaps identified by Bakker have been filled, there will continue to be uncertainty about the proposed association of securitization of water with water security. All too many analysts conflate the first and third concepts as described above. Others propose definitions that may be appropriate for the individual but inadequate for the community, region or nation-state. Of course, every state is obligated to provide the necessities of life to its citizens if it expects to remain in power. However, riots that threaten the longevity of a regime seem to arise more often from increases in the price of bread or petrol than from lack of water. Nevertheless, the obligation is there, and it does apply to water as well as other necessities; and to that extent at least, the obligation is essentially equivalent to the concern for water security in other contexts. The questions are just how the state meets those needs, and whether (or for how long) it can survive with second-best or lower-quality supplies, particularly in urban areas.

In an article about the meaning of "reasonable and sustainable", Wolf (1999) suggested that its operational implications are better explored by *ex post* reviews of current agreements for sharing water than by *ex ante* definitions of what it should mean. The same is probably true of "water security". We lack a well-documented and analytical review of what it has meant up to now. However, that gap need not pose a barrier to indicating how securitization of water as commonly understood erects barriers to more efficient, equitable, sustainable and implementable management of transboundary water, which is the subject of the next section.

Failures of securitized and fixed approaches to sharing water

Any agreement on sharing water across national borders must avoid what has been called "hydrocentricity" (Brichieri-Colombi, 2004) or "hydrocracy" (Molle, Mollinga, & Wester, 2009). Both terms denote excessive emphasis on large-scale water infrastructure, such as high dams, and on policy that places water at the centre of state building. Sensational reporting sometimes presents fresh water as the key to national security, but that is rarely if ever the case. None of the recent books on the Israeli–Palestinian conflict (for example, Caplan, 2009; Gorenberg, 2011; Ross & Makovski, 2010) gives more than minor attention to the role of fresh water as a decisive issue between the two sides. Only Heller and Nusseibeh (2001) devoted more than a paragraph to water, and even their chapter merely says that the Palestinian community needs more water than it currently gets and that sharing water is preferable to unilateral action, neither of which points is seriously in question. As these examples demonstrate, water is *not* a major issue in the Israeli–Palestinian conflict.

Defects of quantitative ways of sharing water

The common approach to sharing water involves treating a water body as a pie. The available water is divided among the riparian nations by a quantitative formula involving absolute or percentage shares. This was the approach used for the Israel–Jordan Peace Treaty and for the Oslo agreements; it is currently reflected in proposals put forward by both the Israeli and the Palestinian negotiating teams (Lautze & Kirshen, 2009; Lautze, Reeves, Vega, & Kirshen, 2005). Unfortunately, though this once-and-forever approach can resolve some short-term issues, it is seriously deficient in the longer term.

Quantitative approaches to sharing water have three defects: securitization, rigidity, and ecological fiction.

As indicated above, water becomes "securitized" when it is portrayed as an essential component of national security. It then leaves the realm of what is negotiable and subject to compromise (Trottier, 1999; Trottier, 2008; Zeitoun, 2007). Any change in existing allocations is perceived as a threat to national security, and proposals for change are portrayed as treasonous.

Quantification also leads to rigidity. As a result of climate change, renewable water resources are likely to decrease in the Middle East, with particularly severe effects on agriculture (FAO, 2008; Freimuth, Bromberg, Meyher, & Al-Khatib, 2007). Quantitative allocations that are possible today may become impossible in a few years. Population growth and economic development can also challenge fixed allocations of water. Larger populations almost always require more water, but changing economic structures can increase or decrease needs for water. Palestinian farmers in the West Bank would pay the current Israeli price for greater quantities of water than they receive at this time. However, the Palestinian economy is likely to turn gradually toward high technology, which will mean a lower demand for water in the future, even if population and incomes increase.

Finally, quantification implies that water can be treated as stable in space and time, when in reality water is inherently mobile in space and variable over time. Water is used over and over again between the time it falls as precipitation and the time it "disappears" into the sea or a deep aquifer or evaporates into the atmosphere. With some exceptions, water quality tends to decline with each use. And in many cases, changes of use imply management by different bodies with different objectives and different rules for use, reuse and disposal. To complicate the situation, semi-arid areas are subject to wide variations in rainfall from year to year, so planning must focus on extremes rather than on averages.

Scientists, experts and citizens

The need to avoid quantified formulae for reaching political decisions about sharing water requires a different role for scientists from what they currently have (Callon, 2003). The problem is not that the views of scientists are irrelevant; on the contrary, scientific analysis is essential to good policy making. Rather, the problem is that science is not so objective as commonly assumed. William Ophuls, an ecological and political philosopher, has expressed the problem elegantly (2011, pp. 134–135):

> Science may indeed provide us with true opinion concerning certain aspects of human nature and the natural world so that we can choose a rule of life that does not flout reality. But it cannot tell us what reality ultimately is, and it cannot choose the rule for us.

David Schindler, one of Canada's most honoured scientists, is less elegant but more direct (Bocking, 2012, p. 23, emphasis in the original):

> I certainly wouldn't want scientists setting policy – I think it would be a disaster – but I think science should be done and well-considered *before* policy is made, and it should be a part of the decision.

Scientists have a critical role to play in identifying options for water management, but their role as scientists stops short of determining which is the best option. Political, social, economic, and environmental values differ between and within societies, and those differences lead to different preferences for how water is used (Linton, 2010; Molle, 2009; Trottier & Brooks, 2013). Those arrangements must also respond to the conventional criteria for efficiency, equity and sustainability – none of which is uniquely definable. And to ensure implementability, they must also recognize the social capital of the people or peoples who will have to implement its provisions. In the Israeli–Palestinian case, this last issue means, among other things, allowing for the largely top-down, centralized means of water management in Israel, and for the largely bottom-up, communal means of water management in Palestine (Trottier, 1999). Each type of management has advantages and disadvantages, and those advantages and disadvantages affect different groups differently. At least in the Israeli–Palestinian case, disciplinary differences outweigh national differences in disputes about proposals for water management and governance (Trottier & Brooks, 2013).

What is required is a more nuanced and more flexible management structure for water that is shared by two or more states. As stated by Varis, Biswas, and Tortajada (2008, p. xi):

> Treaties and institutional arrangements cannot remain static. Factors like water requirements, use patterns and efficiency of management change with time, as do water management paradigms, practices and processes.... It may not be an easy task to formulate dynamic treaties, but one that must be considered very seriously in the coming years.

The next section presents a possible institutional structure for Israel and Palestine that is designed to exhibit the characteristics suggested by Varis, Biswas and Tortajada and simultaneously to avoid the securitization of water resources. Though designed for Israel and Palestine, there is nothing in that structure that is unique to their water resources or political situation. Rather, they serve as an example of how to respond to the need for greater attention to the socio-political aspects of resource development, as recommended by Bleischwitz, Bahn-Walkowiak, Ekardt, Feldt, and Fuhr (2012). They can therefore can be adapted for any region where joint management is needed but in dispute.

An institutional structure without quantified allocations of shared water

The institutional structure proposed for joint Israeli–Palestinian management of shared water is shown in Figure 1. This structure accepts the general principles of shared water management as they have developed in recent years (Kliot et al., 2001; Rahaman, 2009; for groundwater, see the entire September 2011 issue of *Water International*) but goes beyond them to emphasize equality in all rights and responsibilities related to joint management of shared water. Equality does not mean that each side will receive an equal volume of water. It does mean that each will have equal standing within the institutions for joint management and equal opportunity to participate in decision-making processes, criteria that imply that it is not so much water but management that is really shared. Space limitations permit only a brief discussion of the ways in which these bodies would operate; see Brooks and Trottier (2012) and Brooks et al. (2013) for details.

The main tools in the hands of the institutions shown in Figure 1 are continuous monitoring and ongoing mediation, together with information from two reporting agencies. One of these agencies provides interdisciplinary scientific advice; the other expresses community concerns and reflects social capital. Far from being simply a *modus operandi*, the combination of monitoring and mediation is the basis on which decisions will be reached concerning withdrawals from each well, reservoir or spring. The key agencies in Figure 1 are the Bilateral Water Commission (BWC) and the Water Mediation Board (WMB). The BWC will have responsibility for all shared water (not just Palestinian water, as is the case with the current Joint Water Committee). It has a mandate to establish limits for withdrawals, standards for treatment, and targets for releases of water on the basis of recommendations from its advisory bodies.

The decisions of the BWC must balance demands for economic and social development with the imperative to avoid degradation or depletion of the limited supplies of fresh water. Controversies are bound to arise, and that is where the WMB comes into play. It will receive the complaints of any community or body that argues that it is being negatively affected by either a planned water project or an ongoing practice within another

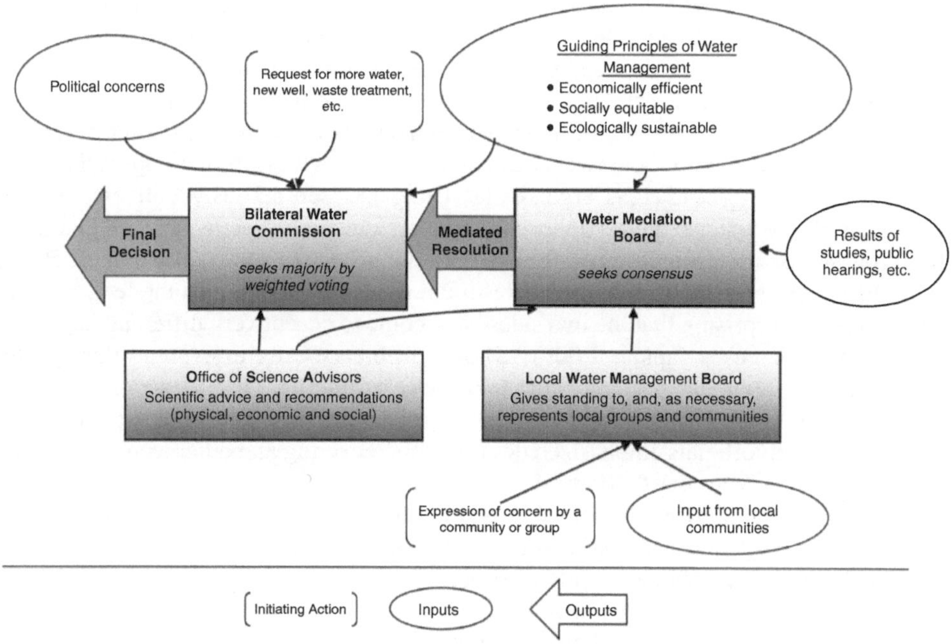

Figure 1. Flow chart of activities and decision making for water shared by Israel and Palestine. Source: Brooks and Trottier (2012).

community or body. It will also receive complaints related to inequitable distribution of water, inadequate water quality, or insufficient attention to environmental protection. The WMB has at its disposal a wide variety of tools that it can then use to attempt conciliation and mediation. If conciliation fails or the alleged impact cannot be proven or disproven by the evidence at hand, the WMB will be empowered to investigate independently and reach its own conclusions, which it will convey to the BWC.

The BWC reports to the two governments; it must therefore respond to political processes and be guided by political directives. It does have final decision-making power, but in this model its decisions are highly constrained. Notably, it is proposed that the BWC be empowered to accept or reject, but not to alter, a recommendation from the WMB. In contrast to the BWC, the WMB has no direct link to senior political bodies. It must respond to the concerns of affected groups and organizations and be guided by explicit goals for efficiency, equity, sustainability and implementability. Both the BWC and the WMB must take into account the guidance they receive from their subsidiary bodies. With this balance of power, the process of decision making can go through several cycles until an acceptable agreement is found.

From concept to practice

It is a big step from abstract review of international water management to practical application in specific regions and with specific histories. The institutions described in the previous section are currently being promoted by Friends of the Earth Middle East (FoEME), which stands out among non-governmental organizations for its strength in both research and advocacy in Israel, Palestine and Jordan. The FoEME proposal requires further work to convert the concepts and the institutions outlined above into ongoing processes and operational agencies. However, to some degree it already resembles the Commission format as defined by Lautze, Wegerich, Kazbekov, and Yakubov (2013) for international river basin organizations. Only the Water Mediation Board requires more than a modest amount of innovation compared with agencies already operating in other regions.

There is no guarantee that the FoEME proposal, or any other agreement on water, will be accepted by the Israeli and Palestinian governments. Not surprisingly, after nearly 50 years of military occupation, the two sides come to water with different approaches. For example, Gvirtzman (2012) has argued that Israel has scrupulously delivered to the Palestinians all the water required under the Oslo agreements, whereas Richard and Issac (2012) focused on the difficulties of securing regular access to water on the West Bank, and on the attacks by Israeli settlers on Palestinian water systems. Given the length of the conflict, it is not surprising that the two sides have come to accept very different narratives regarding its sources (Caplan, 2009). Those differences were evident during the multilateral talks on water resources that took place during the Oslo phase of the peace process (Bookmiller & Bookmiller, 1996). Palestinian officials wanted to talk about water rights, while Israeli officials emphasized desalination, recycling and other technologies for increasing supply (Weinthal & Marei, 2002).[2]

Of course, power is by no means balanced between Israel and Palestine, and the status quo on water in particular is highly favourable to Israel (Brooks & Trottier, 2012; World Bank, 2009, especially Chapters 4 and 7; Zeitoun, Eid-Sabbagh, Talhami, & Dajani, 2013). For those reasons, some analysts have concluded that Israeli reductions in withdrawals would be "quite unprecedented" (Lautze & Kirshen, 2009, p. 201). However, most of the burden of reductions in Israeli water use would fall on the agricultural sector (Jägerskog, 2003; Lautze & Kirshen, 2009; Lonergan & Brooks, 1995). Growing environmental consciousness, higher prices for water in both urban and rural areas, and replacement of fresh by reclaimed water for irrigation should mean substantially lower rates of fresh-water use in Israel in future. Israel has a sophisticated economy that is resilient enough to accommodate the ensuing trade-offs and to provide assistance to adversely affected people. As for the Palestinian side, the World Bank (2009) estimates that over 100,000 jobs could be created if more water were available to revitalize its agriculture.

A related criticism is that neither Israel nor Palestine would be prepared to hand over control of water allocations to as-yet-untested bilateral bodies. This concern can be partly answered by suggesting a staged approach to implementation of the proposal. An immediate option for responding would be an Israeli-Palestinian task force to ensure that all communities have at least minimum water supplies (generally taken as 50 to 100 litres per person per day for household use), at affordable prices. Indeed, at a large meeting in East Jerusalem in November 2010 to launch the first version of the FoEME proposal, Palestinian participants made it clear that they would not consider the proposal seriously until immediate inequities in water availability were resolved. A more substantive option would be to implement the FoEME proposal in stages, with the first stage being joint management of the Mountain Aquifer, which has already received considerable discussion (Feitelson & Haddad, 1998, 2000).

Even in the short term, relaxation of the barriers to modern sanitation in Palestine would improve life in Israel because of the improved quality of water flowing across the border. At the simplest level, the FoEME proposal can be seen as a trade: water quantity for the Palestinians and water quality for the Israelis.

Of course, as with any agreement, either side has the power to appoint officials who would act in ways that make the joint bodies dysfunctional. However, experience with the Israeli–Jordanian Peace Treaty of 1994 gives grounds for optimism. In the words of Munther Haddadin (2011, pp. 184–85), former minister of water and irrigation in Jordan:

> [Israelis and Palestinians] realize that conflict would not bring about more water for them but would create a zero-sum game. Conversely, cooperation can yield a positive result from which all parties can benefit.

In the case of Israel and Palestine, and probably other areas where water is in dispute, the greatest barrier may be fear, not of what may emerge from the negotiations but of coming to the negotiating table at all. As Subramanian, Brown, and Wolf (2012) stated, barriers (or, as they prefer, "risks") to reaching agreement on shared water most often stem from perceptions:

> the perception that an act of cooperation will expose the country to harm, will jeopardize something of value to the country, or will threaten the political future of individual policymakers. (p. vii)

They go on to identify five categories of risk. Though elements of all five can be found in Israeli–Palestinian negotiations, the most important barriers seem to fall into two categories:

- *Sovereignty and autonomy*—the Palestinian perception that they have less negotiating capacity and less information than the Israelis; and the Israeli perception that cooperation would intrude into its authority to make independent decisions about national development.
- *Accountability and voice*—the Israeli perception that the Palestinians or the regional institution may not deliver the anticipated benefits; and the Palestinian perception that their interests would not be adequately considered by the Israelis or the regional institution.

Presumably, both Israeli and Palestinian officials also perceive risk as to whether an agreement over water can be sold to the two publics. However, surveys of professionals (Kliot & Shmueli, 1998) have shown wide agreement on the formula for agreement on water, and surveys of the Israeli public shows wide acceptance of the need for such an agreement (Brooks & Trottier, 2012, Annex B). Aaron Wolf (2000) reported that almost all international negotiations over water allocations over the past century or more have proceeded on the basis of each side recognizing the "needs" of the other side(s) rather than on *a priori* principles or rights. The "picnic table summits", which managed water in the Jordan Valley during the nearly 45 years when Israel and Jordan were still nominally at war, show that it is possible to manage shared water (in this case, a border river) jointly, even in difficult political circumstances. In the different context of Australia, Syme, Nancarrow, and McCreddin (1999) reported that in decisions that require balancing the allocation of water between environmental and human uses, local judgments of "fairness" (along with local participation in decision making) play a distinctly greater role than concern for economic efficiency.

In conclusion, though water security is critical to any society, securitizing water (in the sense of asserting a direct link between available water and either military strength or standard of living) only gets in the way of efficient, equitable, sustainable and implementable development of shared water resources. To the contrary, water security in the form of institutions that provide both peoples with adequate water for comfortable lifestyles and productive livelihoods will have to be efficient, equitable, sustainable and implementable if they are not to fail one or the other side, or more likely both sides. And, one might add, if Israelis and Palestinians, can come to agreement over the water they share, there is no reason why the process might not apply elsewhere in the Middle East, and indeed elsewhere in the world.

Notes

1. Formally named the Treaty of Peace Between the State of Israel and the Hashemite Kingdom of Jordan, retrieved from http://mfa.gov.il/mfa/foreignpolicy/peace/guide/pages/israel-jordan%20peace%20treaty.aspx
2. The corresponding author was a member of the Canadian delegation to these talks, and he concurs with the views of Weinthal and Marei.

References

Abramson, A., Tal, A., Becker, N., El-Khateeb, N., Asaf, L. N., Assi, A., & Adar, E. (2010). Stream restoration as a basis for Israeli–Palestinian cooperation: A comparative analysis of two transboundary streams. *International Journal of River Basin Management*, *8*, 39–53.

Axworthy, T., & Sandford, R. (2012). The global water crisis: Framing the issue. In Harriet Biggs (Ed.), *The global water crisis: Addressing an urgent security issue* (pp. 2–7). Papers for the InterAction Council, Hamilton: UNU-INWEH.

Bakker, K. (2012). Water security: Research challenges and opportunities. *Science*, *337*, 914–915.

Bleischwitz, R., Bahn-Walkowiak, B., Ekardt, F., Feldt, H., & Fuhr, L. (2012). *International resource politics: New challenges demanding new governance approaches for a green economy*. Berlin: Heinrich Böll Siftung.

Bocking, S. (2012). Schindler's pissed. *Alternatives Journal*, *38*, 19–23.

Bookmiller, R. J., & Bookmiller, K. N. (1996). Behind the headlines: The multilateral middle east talks. *Current History*, *95*, 35–37.

Brichieri-Colombi, J. S. (2004). Hydrocentricity: A limited approach to achieving food and water security. *Water International*, *29*, 318–328.

Brooks, D. B., & Trottier, J. (2012). *An agreement to share water between Israelis and Palestinians: The FoEME proposal - revised version*. Amman, Bethelehem: Friends of the Earth Middle East. Retrieved from http://foeme.org/uploads/13411307571~%5E$%5E ~ Water_Agreement_FINAL.pdf

Brooks, D. B., Trottier, J., & Doliner, L. H. (2013). Changing the nature of transboundary water agreements: The Israeli-Palestinian case. *Water International*, *38*, 671–686. doi: 10.1080/02508060.2013.810038.

Callon, M. (2003). The increasing involvement of concerned groups in R&D policies: What lessons for public powers? In A. Geuna, A. J. Salter, & W. E. Steinmueller (Eds.), *Science and innovation: Rethinking the rationales for funding and governance* (pp. 30–68). Chelthenham: Edward Elgar.

Caplan, N. (2009). *The Israeli-Palestinian conflict: Contested histories*. Mississauga: Wiley-Blackwell.

Chellaney, B. (2011). *Water: Asia's new battleground*. Washington, DC: Georgetown University Press.

Chenoweth, J. (2008). Minimum water requirement for social and economic development. *Desalination*, *229*, 245–256.

Feitelson, E., & Haddad, M. (1998). *Identification of joint management structures for shared aquifers: A cooperative Palestinian-Israeli effort* World Bank Technical Paper No. 415. Washington, DC: World Bank.

Feitelson, E., & Haddad, M. (Eds.). (2000). *Management of shared groundwater resources: The Israeli-Palestinian case with an international perspective*. Ottawa: International Development Research Centre.

Fischhendler, I., Wolf, A. T., & Eckstein, G. (2012). The role of creative language in addressing political realities: Middle-Eastern water agreements. In S. Megdal, R. Varady, & S. Eden (Eds.), *Aridity, scarcity and shared water resources: Arizona, Israeli and Palestinian perspectives on solving water management challenges* (pp. 1–31). Delft: Delft University Press.

Food and Agriculture Organization (FAO). (2008). *Climate change: Implications for agriculture in the near East*. Rome: Food and Agriculture Organization of the Middle East. Report for 29th FAO Regional Conference for the Near East (NERC/08/INF/5).

Freimuth, L., Bromberg, G., Meyher, M., & Al-Khatib, N. (2007). *Climate change: A new threat to Middle East security*. Bethlehem: Friends of the Earth Middle East.

Gleick, P. H. (2000). How much water is there and whose is it? The world's stocks and flows of water and international river basins. In P. H. Gleick (Ed.), *The world's water 2000–2001: The biennial report on freshwater resources* (pp. 19–38). Washington, DC: Island Press.

Gorenberg, G. (2011). *The unmaking of Israel*. New York, NY: HarperCollins.

Gvirtzman, H. (2012). *The Israeli-Palestinian water conflict: An Israeli perspective*. Israel: The Begin-Sadat Center for Strategic Studies, Bar-Ilan University, Mideast Security and Policy Studies No. 94.

Haddadin, M. J. (2011). Water: Triggering cooperation between former enemies. *Water International*, *36*, 178–185.

Heller, M., & Nusseibeh, S. (2001). *No trumpets, no drums: A two-state settlement of the Israeli-Palestinian conflict*. New York, NY: Hill and Wang.

Jägerskog, A. (2003). *Why states cooperate over shared water: The water negotiations in the Jordan River Basin*. Sweden: Department of Water and Environmental Studies, University of Linköping.

Katz, D. (2011). Hydro-political hyperbole: Examining incentives for overemphasizing the risks of water wars. *Global Environmental Politics*, *11*, 12–35.

Kliot, N., & Shmueli, D. (1998). Real and ideal institutional frameworks for managing the common Arab-Israeli water resources. *Water International*, *23*, 216–226.

Kliot, N., Shmueli, D., & Shamir, U. (2001). Institutions for management of transboundary water resources: Their nature, characteristics and shortcomings. *Water Policy*, *3*, 229–255.

Lautze, J., & Kirshen, P. (2009). Water allocation, climate change, and sustainable water use in Israel/Palestine: The Palestinian position. *Water International*, *34*, 189–203.

Lautze, J., Reeves, M., Vega, R., & Kirshen, P. (2005). Water allocation, climate change, and sustainable peace: The Israeli proposal. *Water International*, *30*, 197–209.

Lautze, J., Wegerich, K., Kazbekov, J., & Yakubov, M. (2013). International river basin organizations: Variation, options and insights. *Water International*, *38*, 30–42.

Linton, J. (2010). *What is water? The history of a modern abstraction*. Vancouver: University of British Columbia Press.

Lonergan, S. C., & Brooks, D. B. (1995). *Watershed: The role of fresh water in the Israeli-Palestinian conflict*. Ottawa: International Development Research Centre. Retrieved from http://web.idrc.ca/en/ev-9325-201-1-DO_TOPIC.html

Molle, F. (2009). Water, politics and river basin governance: Repoliticizing approaches to river basin management. *Water International*, *34*, 62–70.

Molle, F., Mollinga, P. P., & Wester, P. (2009). Hydraulic bureaucracies and the hydraulic mission: Flows of water, flows of power. *Water Alternatives*, *2*, 328–349.

Ophuls, W. (2011). *Plato's revenge: Politics in the age of ecology*. Cambridge, MA: The MIT Press.

Rahaman, M. M. (2009). Shared water-shared opportunities: Associated management principles. *International Water Resources Update*, *22*, 15–19.

Richard, M., & Issac, J. (2012). The water regime in the West Bank. *This Week in Palestine*, (174). Retrieved from http://www.thisweekinpalestine.com/i174/pdfs/article/the_water_regime.pdf

Ross, D., & Makovski, D. (2010). *Myths, illusions and peace: Finding a new direction for America in the Middle East*. New York, NY: Alfred A. Knopf.

Subramanian, A., Brown, B., & Wolf. A. (2012). *Reaching across the waters: Facing the risks of cooperation in international waters*. Water Papers. Washington, DC: The World Bank.

Syme, G. J., Nancarrow, B. E., & McCreddin, J. A. (1999). Defining the components of fairness in the allocation of water to environmental and human uses. *Journal of Environmental Management*, *57*, 51–70.

Trottier, J. (1999). *Hydropolitics in the West Bank and Gaza Strip*. Jerusalem: PASSIA Publications.

Trottier, J. (2008). Water crises: Political construction or physical reality? *Contemporary Politics*, *14*, 197–214.

Trottier, J., & Brooks, D. B. (2013). Academic tribes and transboundary water management: Water in the Israeli-Palestinian peace process. *Science and Diplomacy*, *2*, 60–80. Retrieved from http://www.sciencediplomacy.org/article/2013/academic-tribes-and-transboundary-water-management

Varis, O., Biswas, A. K., & Tortajada, C. (2008). *Preface. Management of transboundary rivers and lakes*. Berlin: Springer-Verlag.

Weinthal, E., & Marei, A. (2002). One resource two visions: The prospects for Israeli-Palestinian water cooperation. *Water International*, *27*, 460–467.

Weinthal, E., Troell, J., & Nakayama, M. (2011). Water and post-conflict peacebuilding: Introduction. *Water International*, *36*, 143–153.

Wolf, A. T. (1998). Conflict and cooperation along international waterways. *Water Policy*, *1*, 251–265, see also Transboundary Freshwater Dispute Database (Oregon State University, Department of Geosciences). Retrieved from http://terra.geo.orst.edu/users/tfdd/

Wolf, A. T. (1999). Criteria for equitable allocations: The heart of international water conflict. *Natural Resources Forum*, *23*, 3–30.

Wolf, A. T. (2000). From rights to needs: Water allocations in international treaties. In E. Feitelson & M. Haddad (Eds.), *Management of shared groundwater resources: The Israeli-Palestinian case with an international perspective* (pp. 27–59). Ottawa: International Development Research Centre.

Wolf, A. T. (2007). Shared waters: Conflict and cooperation. *Annual Review of Environment and Resources*, *32*, 241–269.

World Bank. (2009). *Assessment of restrictions on Palestinian water sector development, sector note for West Bank and Gaza*. Washington, DC: World Bank.

Zeitoun, M. (2007). The conflict vs. cooperation paradox: Fighting over or sharing of Palestinian-Israeli groundwater? *Water International*, *32*, 105–120.

Zeitoun, M., Eid-Sabbagh, K., Talhami, M., & Dajani, M. (2013). Hydro-hegemony in the upper Jordan waterscape: Control and use of the flows. *Water Alternatives*, *6*, 86–106.

Appendix: water resources shared by Israel and Palestine

As shown in Figure A1, the high ground in the land that comprises Israel and Palestine is the divide between drainage westward to the Mediterranean Sea and drainage eastward to the rift valley. The only significant source of surface water in the region is the Jordan River. It has just one major tributary, the Yarmouk River, which joins the Jordan from the east just south of the Sea of Galilee, and which, for part of its course, forms the border between Syria and Jordan.

Ground water is a more important source than surface water in most of the region. Two aquifer systems dominate supply. The Mountain Aquifer is composed of limestone and underlies the ridge that runs from northern Israel through the West Bank and south to about Be'er Sheva. It carries water in three directions, as shown by the arrows in Figure A1. Much of the Mountain Aquifer can be tapped only by wells several hundred metres deep, which means that the water, though high in quality, is expensive. The Coastal Aquifer consists of a series of mainly independent lenses of sandstone. Early orchards began using this water, with the result that quality has declined over time.

All of the streams that rise in the West Bank and cross the Israeli–Palestinian border before discharging into the Mediterranean Sea are shared water, as are the several that rise in Israel and cross the border before discharging into the Jordan River or the Dead Sea. Although most are badly polluted, clean-up of cross-border streams has been suggested as an opportunity for Israeli–Palestinian cooperation (Abramson et al., 2010).

In summary, roughly two-thirds of the fresh water in the Israeli–Palestinian example can be defined as "shared water", with the implication that this water requires some form of joint management. Given the many paths that water can follow between the time it falls as precipitation and the time it evaporates or flows into the sea – and given, further, the many ways that it can be used (and reused) – the designation of water as "shared" is as much a political choice as a hydrological determination. Water resources that are not shared remain subject to the exclusive sovereignty of the state within which they lie.

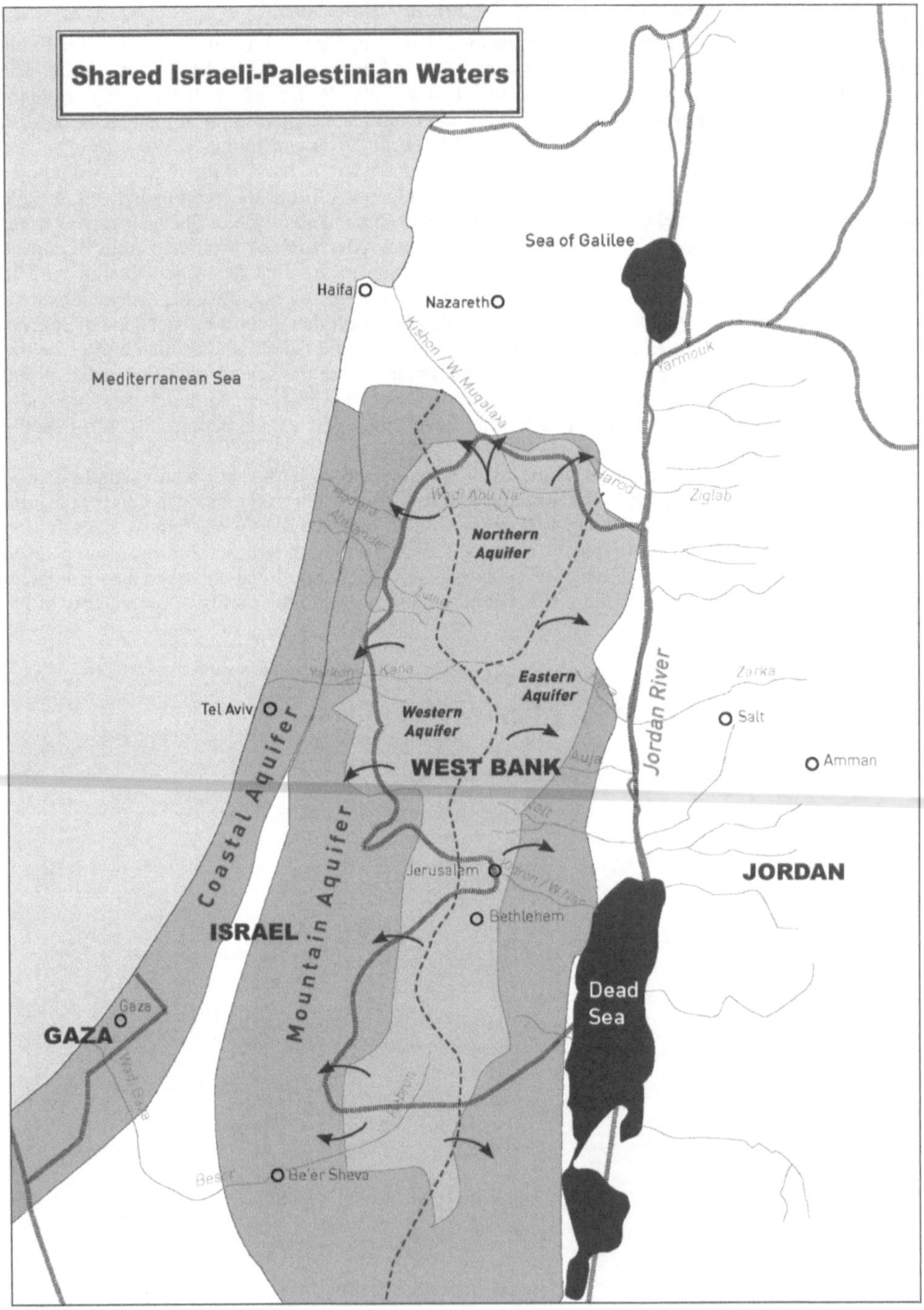

Figure A1. Generalized map of fresh water occurring in Israel and Palestine. Source: Brooks and Trottier (2012).

Index